PHYSIOLOGY OF
THE PEA CROP

PHYSIOLOGY OF THE PEA CROP

Book Coordinated by:

Nathalie Munier-Jolain, Véronique Biarnès,
Isabelle Chaillet, Jérémie Lecoeur,
Marie-Hélène Jeuffroy

With the collaboration of:

Benoît Carrouée, Yves Crozat,
Lydie Guilioni, Isabelle Lejeune,
Bernard Tivoli

CRC Press
Taylor & Francis Group
an **informa** business
www.crcpress.com

6000 Broken Sound Parkway, NW
Suite 300, Boca Raton, FL 33487
270 Madison Avenue
New York, NY 10016
2 Park Square, Milton Park
Abingdon, Oxon OX 14 4RN, UK

Science Publishers
Enfield, New Hampshire

Published by Science Publishers, P.O. Box 699, Enfield, NH 03748, USA
An imprint of Edenbridge Ltd., British Channel Islands

E-mail: _info@scipub.net_
Website: _www.scipub.net_

Marketed and distributed by:

CRC Press
Taylor & Francis Group
an **informa** business
www.crcpress.com

6000 Broken Sound Parkway, NW
Suite 300, Boca Raton, FL 33487
270 Madison Avenue
New York, NY 10016
2 Park Square, Milton Park
Abingdon, Oxon OX 14 4RN, UK

© 2010 reserved

ISBN 978-1-57808-570-5

Published by arrangement with Editions Quae, Versailles

French edition © INRA, ARVALIS-Institut du végétal, UNIP, ENSAM, 2005

ISBN (INRA) 2-7380-1182-9
ISBN (ARVALIS-Institut du végétal) 2-86492-679-2
ISSN 1144-7605

Library of Congress Cataloging-in-Publication Data

Physiology of the pea crop. -- 1st ed.

    p. cm.

  "The French version of this book ([Paris, 2005]) has been coordinated by: Nathalie Munier-Jolain ... with the collaboration
of: Benoît Carrouée ... [et al.]"

Includes bibliographical references and index.

ISBN 978-1-57808-570-5 (hardcover)

1. Peas--Physiology. 2. Peas--Research--France. I. Munier-
Jolain, Nathalie. II. Carrouée, Benoît.

QK495.L52P482 2009

635'.656--dc22

                       2009000071

Printed in the United States of America

Preface

Bertrand Ney and Benoît Carrouée

This book covers all of the applied research carried out in recent years in France on protein peas. It is, however, in our view, more than simple description of progress in agronomic knowledge. It is also testimony to a process of innovation which has involved all the players in the industry. The eighties began with a new challenge for French and European agriculture—to make good the shortages in protein-rich plant products for animal feeding. In effect, soya seed and its cake monopolised the world protein market, whereas the livestock industry was expanding rapidly in France, and demand was growing. The embargo on soya in 1973 and the crisis which resulted in the animal sector quickly showed up the limitations of the European agricultural policy. The « protein plan » intended to promote a French production industry for protein-rich plant products was therefore decided upon in the conditions of the seventies and opened the way to European regulation in 1978. Among the candidate crops was the faba bean, the only protein crop traditionally grown in France for animal feeding, together with soya, pea, and somewhat later, lupin. Soya remained confined to the favourable conditions of the south of France. The white lupin, limited by its sensitivity to calcareous soils, and faba bean, limited by sensitivity to high temperatures, remained confined to the west and north of France. It was peas, hitherto grown in France as a forage for cattle or as a green legume for human food because of its high yield potential and its short growth cycle, which would prove to be the main support of a new industry in protein-rich products not based on oilseed cakes.

From that moment, a new crop would emerge, made possible by the involvement of all the players in the industry: technical institutes (FNAMS, UNIP and ITCF, now called Arvalis-Institut du Végétal), research (INRA) and agronomy schools (Agro Paris Tech, Montpellier SupAgro, ESA Angers, ISA Beauvais etc.), public and private plant breeders, cooperatives and Chambers of Agriculture. Since 1985, these partners have met to confer together within the Pea Agrophysiology Group, on the initiative of UNIP (French Interprofessionnal Organisation of Protein Crops). The results of the first ten years' work were recorded in the first book in 1994 which presented the state of knowledge on the vegetative and reproductive development of protein peas and their use in crop production. Important work by breeders has transformed peas from their former forage architecture (tall plants sensitive to lodging, with a large vegetative yield and an uncertain grain yield) to a stocky and highly branched leafless structure (the leaflets having been replaced by tendrils), which resists to lodging and is easily mechanically harvested, with grain of good size and quality. The crop however remains susceptible to diseases, especially in the north and west, where its abundant vegetative growth creates environmental conditions favourable to disease expression, notably anthracnose (*Mycosphaerella pinodes*), and to water and heat stress in the more continental or southerly regions. Thus it was found, using the first diagnostic tools introduced by the group, that, contrary to recent ideas, a pea crop, although nitrogen fixing, can suffer from nitrogen deficiency under the effects of early water stress, weevils whose larvae destroy the root system, or cultural practices leading to soil compaction. Likewise, its poor root system, in which it is now known that the roots and nodules compete markedly for the products of photosynthesis, accentuates water stress towards the end of growth and can limit yield. These factors, once identified as being liable to harm the crop, have been widely studied during the course of the last ten years and constitute very significant new contributions to this new edition. In spite of its sometimes undesirable characteristics, Solara, a leading variety registered in 1986 and an average ideotype suiting most situations, very rapidly took over the cultivated area. The area sown to peas reached nearly 750,000 ha at the beginning of the nineties. Along similar lines, to promote the crop on the European scale, the AEP (European Association for Grain Legumes Research) was founded in 1992 to bring together all the European players in the industry under the instigation of UNIP.

However, from 1988 and the introduction of budgetary stabilisers, the successive reforms of the Common Agricultural Policy abandoned any will towards targeted development of protein crops : on the contrary, they decided (successfully) to stabilise the areas at the level reached at the end of the eighties in the EU. The pea-growing areas are therefore concentrated in the regions where the climate and the regulations make the crop most competitive. The frequent return of peas to the same fields in these regions during the nineties has led to the development of root diseases, in particular *Aphanomyces*, which causes very serious damage to peas. Many experienced pea growers in these regions have therefore had to abandon the crop for several years to decontaminate the fields, resulting in a marked reduction in the areas.

To cope with these problems, the players in the French industry have decided since the end of the nineties to concentrate on two major target objectives:

- To increase the areas of protein crops by diversifying the crops, with heavy investment in winter peas and faba beans ;
- To develop integrated projects, combining agronomy, genetics, pathology, animal husbandry and economics, in order to develop productive high quality pea ideotypes, more resistant to diseases (especially of the root), and also to evaluate the environmental impact of these innovations.

The agreed efforts for the emergence of this new industry in Europe will not be in vain if one considers the new constraints which will certainly affect the agriculture of the future and the advantages of a crop like peas. This is a good response to the environmental problems and to the necessary diversification of crops for a better control of pathogenic agents, pests and weeds without excessive use of pesticides. The establishment of atmospheric nitrogen fixation from the beginning of crop growth makes the crop independent of nitrogenous fertilisers, which are heavy users of fossil energy and emitters of greenhouse gases. These environmental advantages can only be expressed by assuring optimal management of nitrogenous fertilisation and phytosanitary protection throughout the whole rotation and possibly by introducing a catch crop before or after the peas, whose growing period is very short.

All the work contained in this book follows directly from its predecessor of 1994 and illustrates the dynamism of the partners in agricultural research and development. The new challenges assigned to agriculture demand the efforts of all those involved. The work presented here and what it represents bear clear testimony to this effort.

Contents

PART II: ANALYSIS OF THE EFFECTS OF ABIOTIC AND BIOTIC STRESSES

Preamble: Which approaches can be used to analyze pea canopy physiology?

Jérémie Lecoeur

Choice of approach

Ever since Balls first addressed the question in a study in 1917 (quoted by Evans, 1994) it has long been an objective of researchers to analyze the physiology of vegetal canopies in order to understand the development of crop yields. This objective is as relevant today and it constitutes the framework of this study with reference to field pea.

The approaches used here to analyze the physiology of a pea canopy attempt to define the physical or biotic limits in yield, both quantitatively as well as qualitatively. Over a period of approximately twenty years simple mechanistic approaches, which will be described below, have resulted in significant progress on the fundamental question of pea production limits (Sinclair, 1994). Initially, identification of physical or biotic environmental variables likely to affect crop physiology contributed to this progress, followed by the formalization of the mode of interaction between these environmental variables and the plant. This formalization raised the issue of the mode of representation of the pea canopy adapted to take into account of these interactions between the plant and its environment.

Three approaches were adopted by the scientific and technical community working in the area of field pea productivity:

(i) to favour quantitative and mechanistic approaches, as they appear to be the only alternatives allowing for the analysis and the introduction of the multiple effects of fluctuating and unpredictable environmental conditions on the plant.

(ii) to carry out the analysis of pea crop physiology in two phases;

firstly by describing it outside "biotic and abiotic constraints" in order to establish a reference, then to phase in the effects of different types of constraints. For a field crop plant like pea, analysis of its physiology "without constraint" means, in fact, taking into account only those environmental factors which cannot be controlled by a farmer, i.e. air temperature and incident radiation. For a traditional approach, when these two variables are used, a yield potential afforded by photothermic conditions can be defined regardless of other constraints. The analysis of pea crop physiology under these conditions is the subject of the first part of this paper. The second part deals with the factor-by-factor analysis of the impact of abiotic and biotic constraints taking as a reference the production level afforded by natural photothermic conditions. The third part considers integrating all these results by means of a pea crop physiology model in an attempt to account for field pea yields in fluctuating environmental conditions. The assessment of ground-water, carbon and mass balances of the pea crop allow its environmental impact to be evaluated.

(iii) to move away from traditional plant representations used in field crop models. In the majority of these models, plant representation is limited phenologically to a schedule of crop development stages and, from the plant structure point of view, to a total leaf area, a maximum rooting depth and a biomass. This representation type is suitable for crops such as wheat, corn and sunflowers, as these plant structures are relatively stable from one situation to another. Plant size can vary significantly between situations, but the allometries within the plant are generally maintained. In the case of a grain legume plant with large seeds like the field pea, two physiological specificities have resulted in this mode of representation being modified considerably: indeterminate development and symbiotic fixation. The variability of the number of organs constituting a plant with indeterminate growth and, therefore, of growth phase durations called for a plant description at organ level. For nitrogen nutrition it was necessary to take into account metabolic pathways having different determinisms and dynamics during the course of the cycle.

A simple and generic analytical framework: the energy approach of biomass production

Seed yield is closely linked to the accumulation of biomass by plants, although in pea there is no consistent one-to-one relationship between these two variables. A high biomass accumulation remains a necessary condition for a high yield, even if it is not sufficient in itself. Biomass production of any crop results primarily from the process of photosynthesis allowing the fixation of carbon-dioxide from the atmosphere by means of solar energy captured by the leaves. The development of production potential depends on:

(1) the quantity of solar energy available,
(2) the capacity of the crop to capture this energy,
(3) the capacity of the crop to transform this energy into biomass,
(4) the conversion of this biomass into seeds.

This proposal of an analytical breakdown of plant physiology is similar to the one used by the majority of crop physiology models. Within this framework, the plant canopy can be compared to a biophysical system capturing light energy via its leaf area, producing biomass from this light energy mainly by photosynthesis and distributing its produced biomass between the various plant organs by means of allocation and remobilisation. By using this analysis, yield (Yd) can be broken down in the following way:

$$Yd = DM \times HI$$

$$DM = \int_{emergence}^{harvest} RUE\ RIE\ PAR_0\ dt$$

with DM, dry matter; HI, harvest index; PAR_0, photosynthetically active incident radiation; RIE, radiation interception efficiency; RUE, conversion efficiency of intercepted radiation into dry matter or radiation use efficiency.

This approach was gradually formalized between the beginning of the 1950s and the end of the 1970s. The main stages of this formalization were notably the defining of the concept of the leaf area index by Watson (1947), then the proposing of an analogy between plant cover and a turbide medium according to Beer's law by Monsi and Saeki (1953, quoted by Evans, 1994) and, finally, the formalizing of the capacity of plant cover to transform intercepted light into biomass by means of the concept of biological efficiency proposed by Monteith (1977).

This formalism presents the advantage of being mathematically simple, while remaining mechanistic. Moreover, each of these terms has a biological significance. The various terms thus refer to broad areas of plant physiology. Radiation interception efficiency depends on the leaf area of the plant, which itself results from the processes of organogenesis, morphogenesis and senescence; biological efficiency is largely a function of photosynthetic activity and yield index is determined by the processes of allocation and remobilisation of the plant reserves.

This approach also helps to clarify interactions between the physical environment and the biological activity of the plant, even though only incidental radiation appears in the equation. It thus emphasizes the importance of solar radiation as a primary energy, essential for all agricultural production. The quantity of solar radiation available on a parcel of land thus represents an environmental resource of the parcel being cultivated which varies considerably from year to year and during the course of a single year. In the context of field crops, the farmer is constrained by the level of this resource. His actions are limited by the extent to which the crop can exploit the resource. Other factors of the abiotic environment are taken into account in relation to their impact on the terms of the equation. The impact of temperature is implicit in relation to its effects on cycle duration and metabolic activity level. Other constraints, such as edaphic water deficit or mineral deficiencies, affect organ development and growth.

Biotic constraints can also be taken into account. They can affect radiation interception efficiency by reducing canopy density, by accelerating leaf senescence or abscission or by filtering light arriving on the limbs. They also reduce radiation use efficiency by limiting photosynthetic activity or by using up the surpluses of assimilates of the host-plant. Their effects on the organogenesis and the morphogenesis of reproductive organs, as well as on nitrogen and carbon fluxes, can also affect the yield index.

In pea a significant yield and variability determinant is the number of seeds carried by the plant (Doré et al., 1998). Moreover, seed weight has an impact on quality. It is apparent, therefore, that a simple seed biomass is not sufficient to analyze pea crop physiology, particularly in response to abiotic or biotic constraints. A combination of "energetics of biomass production" and "yield component" approaches would appear to be necessary in order to carry out an effective analysis of pea cover physiology, especially of its reproductive development.

Plant and pea crop representation mode

The gradual implementation of the approach outlined above led to the proposal of a pea plant representation used in the majority of agrophysiological studies.

One of the major difficulties was to account for the indeterminate development of pea. For this purpose, plant structure was described at phytomer level, i.e. the basic building unit of plant axes. Using this approach it is possible to account for the variations of cycle durations through a reduction of the final number of phytomers per stem, while the development duration of each phytomer remains unchanged.

Another difficulty, in terms of plant representation, was to account for the possible differences in development and growth between plants within the canopy. These differences are primarily the result of ramification. In canopy, the number of ramifications varies from one plant to another. This could have led to an extensive statistical approach of the plant population constituting the canopy. However, in contrast to a plant like wheat, ramification production occurs primarily over a short period at the beginning of the cycle and, under standard canopy conditions, the development and growth of the main stem and ramifications are quite close within a single plant and from plant to plant. This factor led to a simplification being proposed, by considering a pea canopy, not as a collection of plants with a variable number of ramifications, but as a collection of stems having similar development and growth characteristics. This led to the definition of the "average" stem and to the representation of a pea crop as the sum of these "average" stems.

Breaking down crop canopy into "average" stems composed of phytomers has the advantage of allowing the description of the development gradients along the stems. Thus, the collection of phytomers constituting the canopy can be organised into groups sharing the same characteristics, particularly in terms of age and degree of differentiation. Thus, a distinction can be drawn between vegetative and reproductive phytomers. A simplification was proposed for the development of reproductive organs. A reproductive phytomer can carry one or two pods, which in turn can carry several seeds. Although there are development gradients between these pods and seeds, the proposal was to represent all these seeds of a given phytomer by an "average" seed, similarly to the case of stems.

In terms of biomass accumulation, pea plants in a canopy show no specific characteristics. Thus, it was possible to apply the simplest traditional representations using state variables on a m² scale. For light interception, the canopy can be represented by its leaf area or its foliar index (LAI) and a coefficient representing the leaf area efficiency to intercept radiation or extinction coefficient k valid for most varieties. Using this mode of representation, transpiration can also be described at canopy level. For biomass production, an estimation of radiation use efficiency was sufficient in most studies. Roots were generally represented by maximum effective rooting depth and a biomass.

For specific studies, in particular the analysis of metabolic paths or individual organ development and growth, detailed approaches were carried out at tissue level, as in the case of seeds, for example.

Spring peas and winter peas

The field pea was for a long time considered to be a crop cultivated almost exclusively at the end of winter or in spring. When attempts were made to achieve further progress in yield margins, there was renewed interest in autumn sowings which have a theoretical advantage in terms of productivity by extending the cycle length and by avoiding unfavourable climatic conditions at cycle end. New varietal types were included in agrophysiological studies, in particular to determine flowering date, as well as new environmental ranges in order to assess pea physiology during the winter phase. Behavioural analysis of these new varietal types was made possible by the progress achieved in understanding yield development mechanisms in spring pea.

PART I

Pea crop functioning and yield components

1

Vegetative development: The morphogenesis of plant organs

J Lecoeur

Introduction

A pea plant is composed of one or more axes or stems which present the same structural organization. According to the development stages of the plant, organs of different ages and nature coexist on the same stem. In this way, if we take a plant shortly before the end of flowering (fig. 1.1), we will find, ascending from the bottom to the top of a stem, senescent leaves, older less active leaves, active leaves carrying filling seeds in their axil, seeds in embryogenesis phase, and finally, growing young leaves carrying flowers and buds. The stem culminates in a cauline bud comprising the very young microscopic organs, with, at its tip, the cauline meristem.

This acropetal succession, which is made up of organs of different types is a characteristic of plants of indeterminate growth type. The simultaneous presence of these organs results in a sometimes complex physiology due to their multiple interactions. Using research into the development and growth of the pea plant relatively simple physiological principles can be drawn from this complex entity which is described in the course of this chapter.

Organisation of a pea stem

The architectural analysis of plants (Godin *et al.*, 1999) has shown that two concepts have allowed for the complete structural organization of all plants to be described: The concept of axis which covers stems in pea and the concept of phytomer which corresponds to what is commonly referred to as

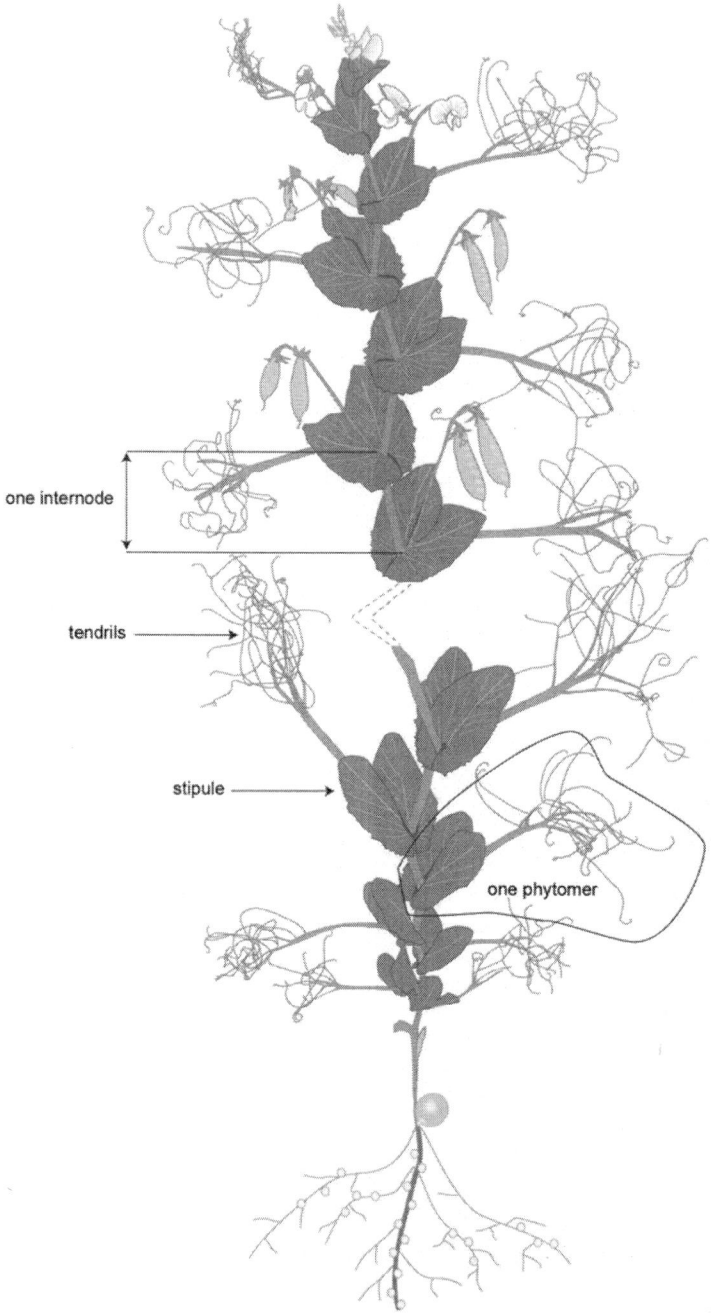

one internode

tendrils

stipule

one phytomer

Figure 1.1. General diagram of a pea plant.

nodes. Following this approach, the axes consist of individual phytomers which have appeared progressively during plant development. These phytomers, which constitute the stem, are all of different ages.

The phytomer acts as the functional unit of the pea plant (Nougarède and Rondet, 1973b). A pea stem consists, therefore, of a collection of phytomers initiated by the cauline meristem. The generic character of the phytomer concept comes from the fact that the organization of a phytomer is identical whatever its age and its position on the stem. It consists of:

— an **internode**, corresponding to the fraction of the stem included between the insertion of the leaf of the previous phytomer and the leaf of the phytomer in question.

— a **leaf**, which has a complex organisation in pea. In this way, in the first pea varieties, known as "leafy", it was made up of a pair of stipules, one to three pairs of leaflets as well as tendrils. In varieties termed "afila", the leaflets are replaced by additional tendrils. There have also been varieties termed "leafless" which have reduced stipules. From an anatomical viewpoint, stipules are not leaves, but mixed folio-cauline organs (Nougarède and Rondet, 1973a). However, from a functional viewpoint, they are considered to be an integral part of the leaf because they represent the largest proportion of transpiring and photosynthetic areas.

— an **axillary meristem** located on the leaf axil. This meristem plays a key role in the development of the plant and its adaptation to its environment. It represents the capacity of the plant to produce new stems, but also to ensure its reproductive development. Thus, a vegetative axillary meristem will generate a new axis or ramification if apical dominance is raised or if conditions for growth are favourable. Ramification presents the same organization as the main stem. If the phytomer carrying the meristem is initiated after flower induction, the axillary will automatically produce a composed inflorescent raceme. In varieties grown in Europe, these racemes comprise one or two flowers depending on radiative and temperature conditions (Hole and Hardwick, 1976).

The cauline bud, beyond the last visible leaf, is composed of approximately ten young phytomers, also of decreasing age, and it culminates in the cauline meristem (fig. 1.2). The cauline bud dies and disappears at the end of flowering with the five to six young organs which did not fully develop. This is the apical stem tip which includes the cauline bud and the unfolding leaf.

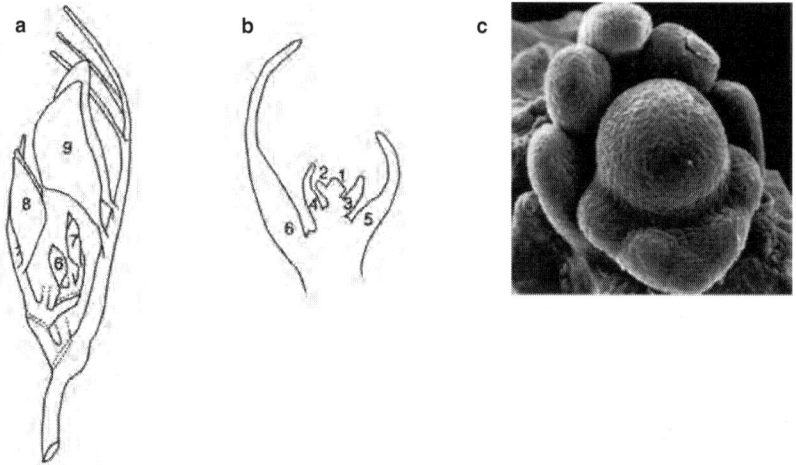

Figure 1.2. Organization of the apical tip. Diagram of the longitudinal axial section of the apical tip of a leafy pea plant before floral initiation and showing the various development stages of phytomers. (a). The four oldest phytomers without their stipula. (b). The last six phytomers and the cauline meristem. The number of each phytomer represents its relative position under the cauline meristem. (c) View of the cauline meristem and the two youngest phytomers (see Lecoeur, 1994; photograph: Nougarède).

Functioning of the cauline meristem and the apical tip

The structural and spatial organization of a pea stem results from the functioning of its cauline meristem (fig. 1.3). This can be described by means of a plant time-scale termed "plastochronic", which is based on the sequential production of phytomers by the meristem. In this way, a plastochrone corresponds *sensu stricto* to all stages separating the initiation of a phytomer from the following one (Nougarède and Rembur, 1985). This scale has been used subsequently to break down the development of the phytomers until maturity into a succession of stages separated by a plastochrone. The plastochronic scale is therefore a differential scale, independent of time, separating the transition between two successive stages. The extension of this approach to the whole development of phytomers has resulted in specific concepts. Thus, the stages separating leaf unfolding from the following stage is called phyllochron, while floral plastochrones refers to those stages separating the appearance of open flowers on two successive phytomers.

The meristem regularly produces phytomer primordia from phytomers, which, by differentiation, will generate internodes and foliar organs. The first visible event in the formation of a new phytomer is the 15° tilt of the meristem in the opposite direction of the medullary zone of the last initiated phytomer. This tilt is due to the reorientation of the mitotic spindles of the medullary zone at the origin of the internodes and, then, to the elongation of the cells of this zone. It always occurs at the same level. This microscopic phenomenon accounts for the alternating spatial pattern of the leaves on the stem with a phyllotaxy of 180°. It gives the characteristic "zigzag" appearance to the pea plant. As it tilts, the meristem grows taller and a meristematic ridge emerges at the base of one of its sides. As it grows, this meristematic ridge soon covers over half of the apical circumference (figs. 2c and 3a). Individual stipula quickly grow out of the sides of this excrescence. The central part of the growth rises forming an emergence from which individual lateral protuberances appear. These will generate leaflets in leafy ones and tendrils in afila (fig. 2c). The final protuberance always gives small tendrils. The individualization of the organs and their differentiation continue in the tip over several plastochrones (Lecoeur, 1994). The leaf resembles the adult organ in miniature approximately five plastochrones after its neoformation. It continues to lengthen for approximately five plastochrones while still folded before it emerges from the apical bud. The unfolding of the leaf out of the apical tip has been the subject of extensive research. Maurer *et al.* (1966) proposed a system of decimal notation (from 0 to 1) which describes precisely the portion of leaf development included between the point it emerges from the apical tip until the point it finishes unfolding. This notation allows for a continuous description of foliar development at plant level and a statistical approach to development at canopy level with a high temporal degree of accuracy.

Modelling foliar development

In many species, plant development rate has been linked to environmental conditions by calculating indicators which take into account temperature, radiation or day-length (Rickman and Klepper, 1995).

• Modelling using the thermal time

The foliar development of a phytomer is delimited by two stages; phytomer initiation by the cauline meristem and the end of foliar organ expansion. The change over the cycle of the number of initiated phytomers and leaves having finished their expansion on a stem was correlated with average air temperature (Truong and Duthion, 1993; Turc and Lecoeur, 1997). This correlation was formalized with thermal time expressed in degree-day units, obtained by adding temperatures above a base temperature (Bonhomme, 2000). The most commonly used base temperature for the calculation of thermal time in pea is 0°C (Ney and Turc, 1993).

The change in the numbers of initiated and expanded phytomeres with thermal time from emergence (fig. 1.3) allows to propose a diagrammatic spatial and temporal representation of the foliar development of phytomers of a pea stem known as the foliar development model (Turc and Lecoeur, 1997). The plastochrone and the phyllochron correspond to the time separating initiation or the end of expansion of two successive phytomers. They are constant when they are expressed in thermal times for different groups of phytomers (fig. 1.3):

— the plastochrone of the phytomers until position 19 on the stem is 40 % shorter than the plastochrone of the following phytomers,
— the phyllochron of the phytomers in positions one to six is 15 % longer in comparison with the following ones.

In the absence of abiotic stresses, these values are independent of the trophic state of the plant for a very wide range of growing conditions including varying canopy densities and differing levels of incident solar radiation (Turc and Lecoeur, 1997). Identical values of plastochrones and phyllochrons were found for a wide range of genotypes including leafy plants and afila as well as spring and winter peas. However, variations of phyllochrones have been observed in a large range of genotypes (Bourion *et al.*, 2002).

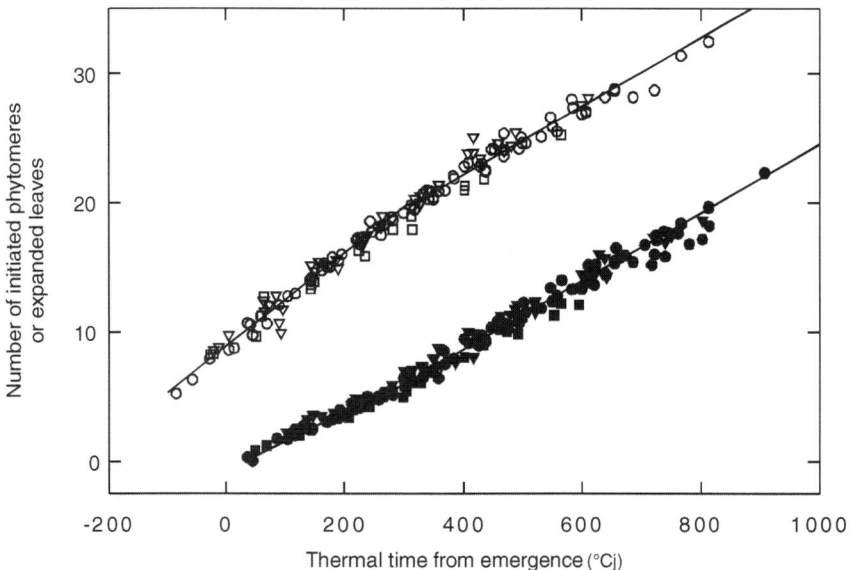

Figure 1.3. Diagram of the foliar development model of a pea stem. Changes in the number of initiated phytomers (open symbols) and of the number of leaves having completed their expansion (full symbols) on a stem in relation to thermal time since emergence (temperature base: 0°C). Atol, Alex, Baccara, Frilène, Messire and Solara varieties. The different symbols represent different experiments (adopted from Turc and Lecoeur, 1997).

- ### Foliar development from phytomere initiation to the end of leaf expansion

This diagram shows that:

— approximately five initiated phytomers in the embryonic seed axis and eight to nine at emergence;

— we should make a distinction between two phases in foliar stem development, ie. between the emergence and up to 300°Cj from emergence, the plastochrone is greater than the phyllochrone with that of appearance of the leaves having finished their expansion. Next, two rates are identical and remain constant as long as seed development does not occur. The result of these coordinated changes in the plastochrone and the phyllochron is that the number of phytomers in the apical tip increases from nine to thirteen during the first 300°Cj of the cycle, and stabilizes subsequently;

— the development duration of the leaves increases for the phytomers in position 1 to 19. It then remains stable to reach approximately 600°Cj, i.e. longer than a month from leaf initiation. The great stability of this model in the absence of edaphic water constraints and mineral deficiency allows a precise estimation of the number of initiated organs and number of leaves having finished their expansion by using average air temperatures. These results can be simply linked to the notations of leaf unfolding on Maurer *et al.*'s scale (1966). A leaf has finished its expansion when the following leaf has unfolded to 0.6 on the scale.

- ### The leaf from the unfolded stage to the senescence

The development model can be extended to stages beyond the end of leaf unfolding and expansion (Pic, 1998). Thus, other leaf development stages which occur on stable thermal dates after leaf unfolding can be added to the stages which have been described previously: during the 250°Cj which follow leaf unfolding, the leaf experiences a maturation phase during which photosynthesis increases gradually until reaching its maximum. During this phase, ARNm content falls by 90%, proteins content falls by approximately 30%, whereas RubisCo and chlorophyll content more or less maintain their level at the time of unfolding; the leaf then has a period of optimal photosynthetic functioning of approximately 200°Cj; then the leaf enters senescence gradually. Its RubisCo and chlorophyll contents decline gradually while there is a build-up of senescence markers such as of cysteine-proteinase ARNm. During this phase photosynthetic activity drops off rapidly. These events occur before leaf yellowing becomes visible. The visual symptoms of foliar senescence are thus not reliable indicators of the leaf's metabolic activity. In the absence of constraints and competition with other organs, a leaf thus has a very long intrinsic lifespan which would be higher than the usual pea cycle duration and which is programmed at molecular level. However, the leaves do not reach this age programmed at molecular level. Leaf senescence, as it is traditionally observed, is not related to the intrinsic age of leaves, but it is triggered at whole plant level, in particular by reproductive development. In this way, all the leaves suffer a drop in their metabolic activity simultaneously. The extent to which this event occurs early on depends on the trophic status of the plant and the environmental conditions. Water deficit accelerates this foliar senescence at plant level, while ablation of part of the reproductive organs delays it (Lecoeur, 1994; Pic *et al.*, 2002).

Expansion of the vegetative organs

Expansion of the organs results from the processes of cellular division and expansion. We have seen that the expansion of a leaf is a long process which lasts several hundreds of degree-days. It can be broken down into two phases:

— a long phase of exponential expansion which represents three-quarters of the total duration of leaf development;
— a short phase of linear expansion (fig. 1.4) (Lecoeur *et al.*, 1995). This last phase corresponds to the visible phase of leaf expansion out of the apical tip. Thus, in the apical tip, out of a dozen young developing phytomers, the eight youngest are in a cellular division phase, while the four older phytomers are in a cellular expansion phase.

- Division then expansion of the cells

The area of a leaf is determined by the area of its epidermis. It is thus the dynamics of division and cellular expansion of this tissue which condition the expansion kinetics of the leaf. During approximately the first two-thirds of leaf expansion, the number of epidermic cells increases exponentially without a corresponding increase in cell size (fig. 1.4). This period corresponds to an active phase of cellular division at a constant rate when expressed in thermal time. Thus, approximately 85°Cj are required for the number of epidermic cells to double, i.e. approximately four days at 20°C. After this phase of cellular proliferation, there is a short transition phase where the rate of cellular divisions slows down, while cell size starts to increase. The number of epidermic cells is determined shortly before the leaf begins to unfold. During the final quarter of its development, leaf expansion

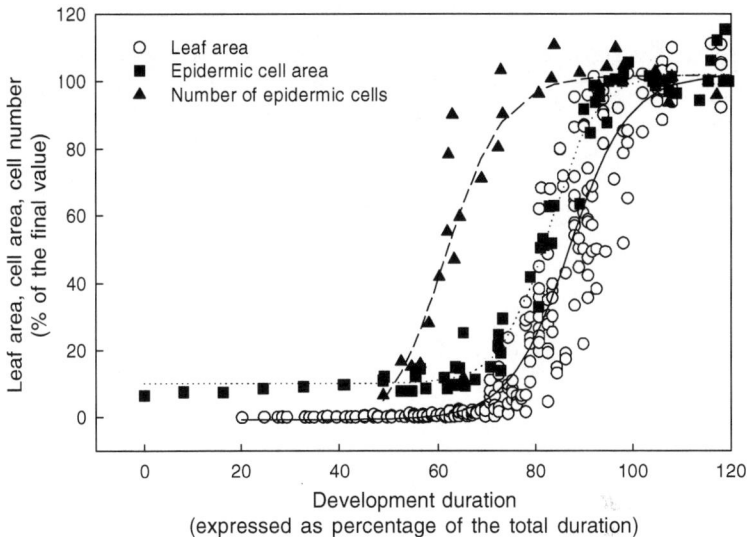

Figure 1.4. **Limb expansion:** Limb area, epidermic cell area and numbers of epidermic cells in relation to time elapsed since phytomer initiation (adopted from Lecoeur *et al.*, 1995).

results from the expansion of cells produced during the previous phase. The other vegetative organs, internodes and tendrils, have the same expansion kinematics. However, they differ from leaves by the spatial organization of their cellular division zones. The cellular divisions of dicotyledone limbs occur simultaneously in all of the zones of the organ, whereas internodal and tendril divisions are confined to a limited meristematic zone at the base of the organ. The expansion mode of these organs is similar to that of the roots and leaves of monocotyledones (Fahn, 1990).

• Final size of vegetative organs

In the absence of water or thermal constraints, the final area of the epidermic cells is stable enough for the phytomers in position four and higher. This means that the differences in leaf area observed between the leaves with different stem positions are a result of a difference in cell numbers. As the division rate is stable in thermal time, it is the differences in phase duration of cellular division which are responsible for the gradual area increase of the leaves with their stem position (Lecoeur and Sinclair, 1996). These differences can be determined by means of the foliar development model. The leaf area profiles of the various vegetative organs always appear in a "bell-shape" of area profiles (fig. 1.5).

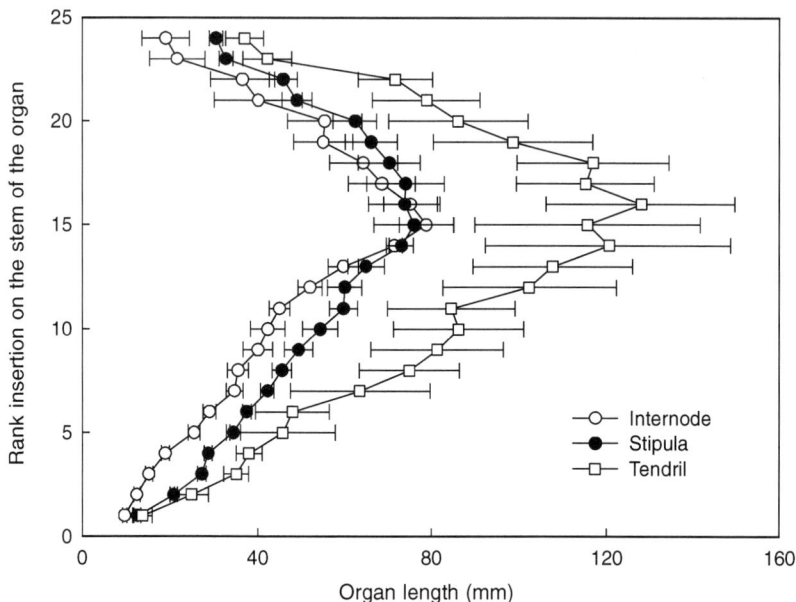

Figure 1.5. Length profiles of different vegetative organs. Internode, stipula and tendril lengths according to position of the corresponding phytomer on the stem. Alex variety, Montpellier experiment 1997 (Lecoeur, np).

The changes in size of the various vegetative organs with their position on the stem are comparable. Size increases for the organs located at the bottom of the stem. It is relatively similar for the organs to be found on the central section of the stem, while size decreases for the organs located at the top of the stem.

We have seen that in the absence of constraint the organ size differences between stem position are related to the duration of the cellular division phase. According to this expansion pattern where the duration of cellular division determines organ size, size should increase until phytomer 19, then remain constant for those following (Turc and Lecoeur, 1997). This organ size profile is obtained in the absence of constraint and competition between the different plant organs. The decreasing size of the organs located on the upper section of the stem is the consequence of the trophic competition between the reproductive and vegetative organs (Lecoeur, 1994). The reproductive organs are large sinks, often having priority, thus limiting the flow of available assimilates for the vegetative organs (see p. 100 and 111). The expansion of the vegetative organs is thus restricted from the point at which the reproductive organs begin to develop. Consequently it can often be noted that the phytomer presenting the largest vegetative organs corresponds to the first reproductive phytomer or is very close to it.

• End of phytomere production

When the demand of the reproductive organs becomes very high in relation to available assimilates, the supply of assimilates at the apical tip ceases. This brings about the end of production and appearance of new phytomers, and leads to the cauline bud dying. Regular ablation of the reproductive organs prevents this competition and allows the production of new phytomers to be maintained (Turc and Lecoeur, 1997).

Conclusion

A pea plant is composed of one or more axes or stems which present the same structural organization.

A stem is made of a succession of structural units called phytomeres, and it ends with the apical tip, which is composed of young phytomere promordia and a cauline meristem. A phytomere includes: an internode, a leaf and an axillary meristem (cf. figs. 1.1 and 1.2).

The plastochrone corresponds to the duration between the production of two successive phytomeres in the apical tip. The phyllochrone corresponds to the duration between the unfolding of two successive leaves on a stem.

The foliar model development proposed is based on a diagram representing of the phytomere initiation and the leaf appearance expressed as a function of cumulated degree-days from emergence, using a 0°C base temperature. This model allows to estimate, at anytime of the plant cycle, the number of initiated and expanding phytomeres in the apical tip and the duration between the initiation and the end of expansion of each leaf.

The size of vegetative organs (internodes, stipula and tendrils) increases from the botton to the middle part of the stem. The phytomere displaying the largest vegetative organs generally corresponds to the first reproductive phytomere or the very next ones. Concerning the upper part of the stem, the size of vegetatives organs decreases because of the trophic competition with the reproductive organs which have priority in carbon allocation.

The rates of cell divisions are constant for all the organs whatever their insertion rank on the stem. The acropetal increases in cell number per organ along the stem is due to the increase of the duration of the cell division phasis. This leads to an increase in organ sizes with their insertion rank on the lower part of the stem.

Floral initiation and the beginning of flowering

Isabelle Lejeune-Hénaut, Véronique Biarnès

Control of the earliness of floral initiation, flowering and maturity is crucial for pea breeders. First of all, the earliness of floral initiation is important because the transition of the apex to the reproductive stage increases plant sensitivity to frost. Secondly, the beginning of flowering is a key stage for the pea crop. It corresponds to a part of the reproductive stage during which the number of seeds is determined, which, to a large extent, determines yield (Doré, 1992). During this period the crop is particularly sensitive to water and temperature stresses (high temperatures) and the effects of these stresses on yield are greater than at other periods of the growth cycle (Lambert & Link, 1958; Salter, 1963; Jeuffroy et al., 1990; Lecoeur, 1994; Ney et al., 1994; Roche et al., 1999). Thus the timing of flowering is a determining factor in final yield. Finally, the earliness of maturity often affects the chances of being able to harvest in favourable climatic conditions. Of course, the three types of earliness we just mentioned are closely linked: generally speaking, if the floral buds are initiated early, flowering and maturity will also be early. Nevertheless, there are variations in this general trend, which we will be discussing in this chapter.

After describing the changes that lead to floral initiation and to the beginning of flowering, we will examine the climatic and genetic factors that determine when these stages occur. We will then assess variability of the length of the period from floral initiation to the beginning of flowering and its consequences in terms of selection, particularly for winter peas.

Definitions and observations

- Floral initiation

It is possible to observe floral initiation of the shoot apical meristem with a binocular magnifying glass (x 40 to x 100), but firstly the most developed leaves have to be removed. In comparison with the vegetative apex (fig. 1.2), the change to the reproductive stage is characterized by the formation of a floral primordium (fig. 1.6), located in the axil of the most recently differentiated leaf. One plastochrone later, the inflorescence is made up of two spheres, each of which will produce a flower.

- Beginning of flowering

A population is considered to be at the beginning of flowering stage when 50 % of the stems display flowers. A stem is considered to be in flower when it displays at least one fully open flower (stage 0.5 on the scale drawn up by Maurer et al., 1966).

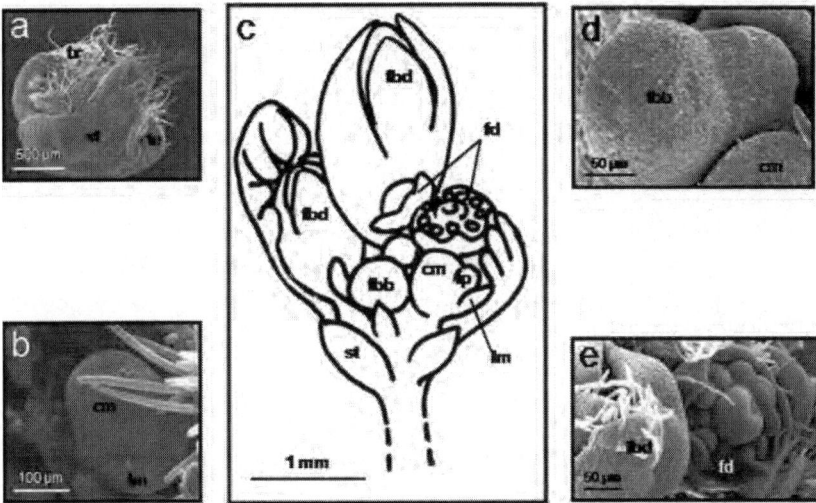

Figure 1.6. Floral initiation of a shoot apical meristem.

a, b, d, e.: scanning electron microscope image (R. Devaux & V. Fontaine, INRA UMR 1281 SADV; G. Vasseur, University of Picardy Jules Verne).

a.: apical bud of a pea stem. The leaf primordia enfold the meristem. As floral initiation approaches, trichomes develop on the stipules.

b.: shoot apical meristem at the vegetative stage. The youngest leaf meristem appears as a ridge at the bottom of the shoot apical meristem.

c.: drawing of a bud after floral initiation. Inflorescences develop rapidly. It is possible to distinguish different stages of development of the inflorescences inside the apical bud well before the appearance of floral buds which become visible during elongation of the last internodes.

d.: inflorescence at the floral bulb stage. The floral primordium is already divided into two meristems that will produce the two flowers of the inflorescence.

e.: floral disc and floral bud. At the floral disc stage, it is possible to distinguish the primordia of the ovary and the 10 stamens surrounded by the sepal primordia. The floral bud stage is reached when the sepals close over the internal organs of the flower.

fbb.: floral bulb; fbd.: floral bud; fd.: floral disc with the primordia of the ovary and anthers; lm: leaf meristem; fp.: floral primordium; st.: stipule; tr.: trichomes; te.: tendrils.

Transition to the reproductive stage in pea

The transition from the vegetative to the reproductive stage occurs on the one hand in response to determining environmental factors, mainly temperature and photoperiod, and on the other hand to internal stimuli linked with development and growth, for example, the number of leaves or the size of the plant (intrinsic earliness). In reaction to these factors, inducer molecules are transported from the leaves to the apical meristems of the stems, after which the developmental events that characterize floral initiation can occur, if the inducing molecules reach a bud which is competent for flowering.

The network of signals that leads to floral transition is determined by the genetic identity of the plant. In pea, Murfet *et al.* conducted a large number of genetic and physiological studies of floral mutants and their offsprings. These authors identified a system of six interactive genes (*Lf, Sn, Dne, Ppd, E* and *Hr*) that determine the first flowering node as well as the dates of floral initiation and flowering (Murfet, 1971, 1973, 1975, 1985; Arumingtyas & Murfet, 1994; Weller *et al.*, 1997). The five loci *Sn, Dne, Ppd, E* and *Hr* control the sensitivity of pea to temperature and photoperiod. The *Lf* locus, for which four different alleles have been identified (Murfet, 1975), controls the intrinsic earliness of each genotype. *Lf* controls the level of receptivity of the apex to the flowering signal: varieties differ in the length of exposure to the flowering signal required by the apical meristem in order to switch to the reproductive stage. In other words, differences in earliness observed even when the photoperiod is optimal are due to allelic variability at the *Lf* locus. Only the *dne* and *ppd* alleles correspond to induced mutations, the four other genes have been described through natural mutations and the behaviours described by Murfet can be observed with relative ease in cultivated varieties. Along with genetic studies, the position of these main flowering genes on the genetic map of pea has been determined and overviews are available (see review on flowering genes in Murfet and Reid, 1993 and pea consensus maps in Weeden *et al.*, 1998 and Ellis and Poyser, 2002). Moreover, the development of molecular tools has allowed to identify molecular markers tightly linked to some loci (*Sn* and *Dne*: Rameau *et al.*, 1998; *Hr*: Lejeune-Hénaut *et al.*, 2008) and sometimes to determine the gene coding sequence (*Lf*: Foucher *et al.*, 2003).

In the paragraphs that follow, we give the main results of physiological studies that characterized the transition to the reproductive stage and establish links with elements of characterization of the main flowering genes known today.

• Effects of photoperiod

Simultaneously with studies on floral mutants, certain authors reported the effects of photoperiod and temperature on cultivated varieties of pea. The first results of these studies highlighted the accelerating effect of long days on the transition to the reproductive stage (Barber, 1959; Aitken, 1978; Berry & Aitken, 1979; Truong & Duthion, 1993; Lejeune-Hénaut *et al.*, 1999). As proposed in genetic studies by Murfet, the main characters reported to describe floral transition are the date of floral initiation and the beginning of flowering, and the position of the first flowering node.

In the majority of varieties studied, and in particular in European dry pea cultivars, a quantitative effect of day length can be observed on the transition to the reproductive stage. Thus the longer the daylength experienced by the plants after emergence, the earlier the floral initiation will begin. The amplitude of the reaction varies with the variety. For instance, Lejeune-Hénaut *et al.*, (1999) showed that when sown in the field in autumn (short days) the Solara variety required 100 degree-days more than when sown in spring (long days) to reach floral initiation. The additional time required can reach 200 degree-days in a variety like Kazar. Berry and Aitken (1979), reported delays of around 25 days in the beginning of flowering at 18°C for the varieties Collegian and Greenfeast when cultivated under a photoperiod of 8 h instead of 16 h. Comparison with genetic studies by Barber (1959), Murfet (1971, 1985), Arumingtyas & Murfet (1994), showed that the dominant alleles *Sn, Dne* and *Ppd*, which constitute the system that controls the production of a floral inhibitor, are responsible for this quantitative response to the photoperiod.

However, within the limits of genetic variability of pea, qualitative reactions to photoperiod do exist.

Certain varieties, such as the American cultivar Alaska (Aitken, 1978), are aphotoperiodic, i.e. they initiate and develop flowers independently from daylength, retaining the same position of the first flowering node. Genetic studies by Murfet (1971a, 1971 b) showed that this behaviour is linked to the recessive *sn* mutation.

Other varieties have strict photoperiodic requirements. This is the case of certain fodder peas (fig. 1.7; Lejeune-Hénaut, 1999). Even when these varieties are sown before October 15, floral initiation does not occur until daylength reaches a threshold of 13.5 hours, i.e. around April 15 of the following year at a latitude equivalent to that of northern France. Comparison of this response with data in the literature enables us to state that, in its dominant form, the *Hr* (*High response*) gene is mainly responsible for plants remaining in the vegetative stage under the effect of short days. *Hr* acts as an amplifier of the *Sn Dne Ppd* system (Murfet, 1973).

• Effects of temperature

Several authors also showed the influence of temperature on the occurrence of floral initiation and flowering. These authors all underline a decrease of the first flowering node position under the influence of low temperatures, corresponding to a vernalization effect. Moore and Bonde (1962) reported that vernalization occurs at temperatures between 4°C and 8°C. From a genetic point of view, Barber (1959), and Murfet (1973) showed that the dominant alleles *Sn, Dne* and *Hr* enable the plant to respond to vernalization in addition to photoperiod.

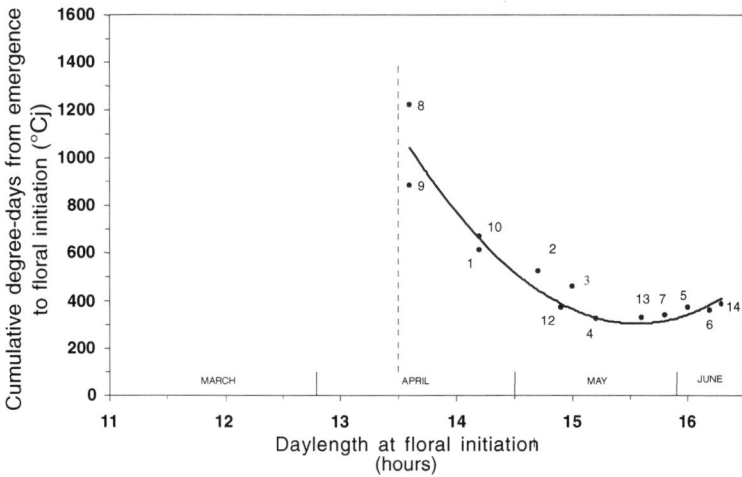

Figure 1.7. Reactivity of fodder pea to photoperiod (after Lejeune-Hénaut *et al.*, 1999; © Kluwer Academic Publishers, 1999; reproduced with the kind agreement of Springer Science and Business Media).

Data were obtained for 14 sowing dates in the field at INRA, Mons-Péronne, France, between November 1995 and May 1997 (**1:** November 2, 1995; **2:** November 27, 1995; **3:** December 18, 1995; **4:** March 14, 1996; **5:** April 15, 1996; **6:** May 13, 1996; **7:** June 11, 1996; **8:** September 17, 1996; **9:** October 14, 1996; **10:** November 8, 1996; **11:** February 10, 1997, no data available for Champagne at this date; **12:** March 11, 1997; **13:** April 10, 1997; **14:** May 26, 1997).

The duration of the period from plant emergence to floral initiation is expressed in cumulative degree-days, base temperature 0°C. The photoperiod given for the day of floral initiation corresponds to the interval between sunrise and sunset at the Mons-Péronne site.

• Effects of interaction between photoperiod and temperature

The main effects of photoperiod and temperature on the transition to the reproductive stage are quite well known. However, significant interactions also exist between the two factors. Berry & Aitken (1979) reported that the varieties Collegian and Greenfeast did not respond to photoperiod at a temperature of 6°C for floral initiation and the beginning of flowering, but displayed a photoperiodic response at 12°C, 18°C and 24°C. This interaction effect is even more marked in the Dun and Mackay genotypes. The authors attribute this variability of response to the temperature x photoperiod interaction to the *Sn hr* allelic combination for the first two varieties comparatively with *Sm Hr* for the second two.

Modelling "floral initiation" and "beginning of flowering"

In natural conditions, where there is constant variation in temperature and photoperiod between sowing and floral initiation, it is difficult to integrate the major effects of and interactions between the two climatic parameters. Certain authors have nevertheless proposed models for the position of the first initiated flowering node (Truong & Duthion, 1993), the date of floral initiation (Lejeune-Hénaut et al., 1999), and the date of the beginning of flowering (Summerfield & Roberts, 1988; Truong & Duthion, 1993).

For the beginning of flowering, most spring genotypes respect the equation proposed by Summerfield and Roberts (1988), which takes the form $1/f = aT + bP + c$, and links the date of beginning of flowering (f) with temperature (T) and photoperiod (P). This type of model which integrates the photopheriod, comparatively with a model simply based on cumulated temperature after sowing, allows forecast of the flowering date to be extended to very early sowing (January) and very late sowing (April/May) as demonstrated by Biarnès-Dumoulin et al. (1998). According to Roche et al. (1999), taking latitude, sowing date and photoperiod into account appears to be the optimal model for predicating the date of flowering in the variety Solara.

Using a mechanistic approach, Truong & Duthion (1993) showed that the date of beginning of flowering depended both on the level of the first flowering node to be initiated and on the rate of leaf appearance (phyllochron, see also p. 6). In this type of model—which takes the rate of development into account—the variability of the date of beginning of flowering mainly depends on the temperature and the photoperiod which determine the first flowering node to be initiated, and then on environmental conditions which influence the rate of leaf appearance. In the absence of abiotic stresses and under a very wide range of growth conditions, temperature is the main factor that influences the rate of leaf appearance expressed in days (see p. 7 and 8). Only severe hydric stress can lead to a significant change in the rate of leaf appearance (see p. 140).

Use of flowering date to identify genes responsible for the reaction to photoperiod

The qualitative response of fodder peas (Hr) to photoperiod is of particular interest in the creation of winter pea varieties. The fact that these genotypes remain in a vegetative state enables them to survive major periods of winter frost. During autumn, the plants develop several branches but remain dwarf (short inter-nodes, reduced stipules and leaflets) with a creeping growth habit; they retain this "rosette" habit throughout winter and only start developing again in spring. From the development point of view, this particular morphology is linked to the fact that floral initiation, which signals the beginning of a period of increased plant sensitivity to frost, is delayed to a period when climatic conditions are more favourable.

Identifying *Hr* genotypes is thus particularly important in programmes for improvement of frost resistance in pea. As it is laborious to dissect a large number of apices, the date of the beginning of flowering can be used to make a preliminary screening.

In this way, sowing the same material at different periods, for example one sowing in autumn and one in spring, enables to observe different behaviours between varieties with respect to photoperiod. Figure 1.8 shows an example of varieties sown in Maisse (Essonne, France) in October 2001 and in March 2002. *Hr* type varieties like Champagne display little variation in the date of the beginning of flowering expressed in calandar days: irrespective of the sowing date, floral initiation in Champagne did not occur

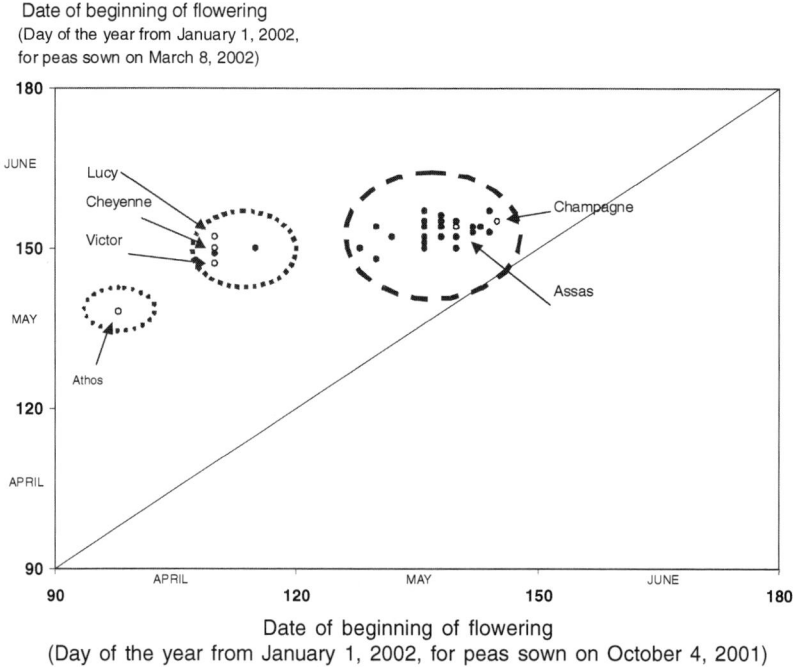

Figure 1.8. Comparison of dates of beginning of flowering after autumn and spring sowings showing different types of response to photoperiod (trials by *GAE Recherche*, 2001–2002). Data were obtained for two field sowing dates (October 4, 2001 and March 8, 2002) at *GAE Recherche* in Maisse (France). Varieties are indicated by empty circles (○) and *GAE Recherche* lines by full circles (●).

before daylength reached 13.5 hours (see p. 17, fig. 1.7), which implied flowering began between May 24 and June 4. As an initial approximation, it is possible to extrapolate this type of reaction to all the lines close to the first bisector in the graph. In lines that do not respond to photoperiod (*hr*) like Cheyenne or Athos, the flowering date is mainly determined by the sum of degree-days after sowing, and peas sown in October thus flower earlier (in calendar days after January 1) than peas sown in March, although with differences linked to their respective intrinsic earliness (*Lf* gene).

Genetic variability of the length of the period from floral initiation to the beginning of flowering

In the case of winter peas, the objective of achieving late floral initiation but early flowering may at first seem contradictory. However, the available allelic variability for *Hr* and *Lf* genes makes it possible to envisage a winter pea with:

— a delayed floral initiation to avoid frost, using the dominant *Hr* allele;
— a relatively early flowering date to ensure seeds develop when climatic conditions are optimal, using the allelic variability available at the *Lf* locus.

This particular allelic combination is made possible by the fact that the two loci are located on two different chromosomes and are genetically independant. This hypothesis is also supported by evidence for genetical variability for the period from plant emergence to floral initiation (Berry & Aitken, 1979; Lejeune-Hénaut, unpublished). Figure 1.9 shows that for the same date of floral initiation, the difference in the beginning of flowering between genotypes can reach 300 degree-days. Recent sequencing of the *Lf* gene should enable the development of molecular tools to facilitate the introgression of favourable alleles at the *Hr* and *Lf* loci into lines of agronomic interest.

Conclusion

The transition from the vegetative to the reproductive stage (floral initiation and beginning of flowering) is a significant event in the development cycle because it plays a determining role in the plant's ability to adapt to environmental conditions. In pea, considerable knowledge has been acquired in different domains:

— in physiology: it has been established that a flowering signal— whose nature is not yet understood—is produced by the leaves in response to photoperiod, temperature, and the age of the plant; this signal is transmitted by the sap to the terminal meristem and triggers floral initiation;

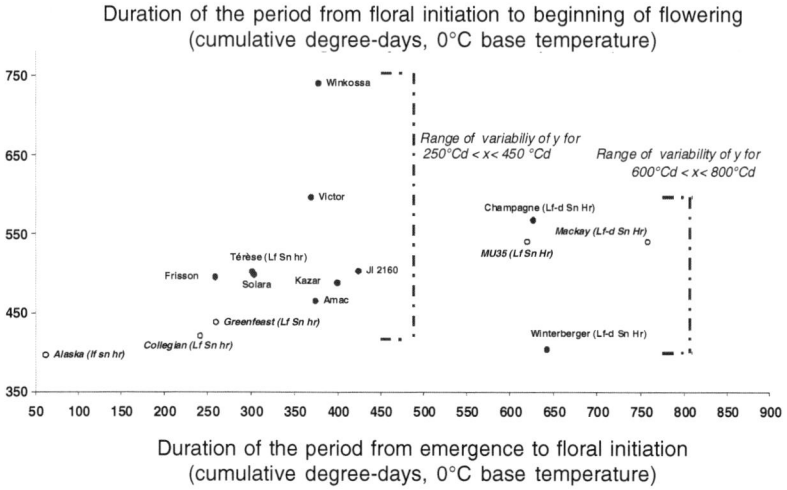

Figure 1.9. Duration of the "floral initiation—beginning of flowering" period as a function of the duration of the "emergence – floral initiation" period (after Berry & Aitken, 1979, and Lempereur, 1996). For the varieties Alaska, Collegian, Greenfeast, Mackay and the line MU35 (empty circles o, labels in italics) the coordinates were calculated following observations made by Berry & Aitken (1979) in controlled temperature conditions (6°C) and with a controlled photoperiod (8 h). Among the different conditions tested by these authors, those presented here are those that best corresponded to the field conditions used to evaluate the varieties Amac, Champagne, Frisson, Kazar, Solora, Térèse, Victor, Winkossa and Winterberger and the line JI 2160 (full circles •, mean temperature and photoperiod between plant emergence and floral initiation at around 3°C and 12 h respectively for peas sown on November 2, 1995).

The genotype of the different lines for the *Lf Sn* and *Hr* loci is shown in brackets (when known). As can be seen, there is a wide range of variability for the duration of the "floral initiation-beginning of flowering" period irrespective of the earliness of floral initiation. In other words, it is possible to find varieties with late floral initiation (group of *Hr* lines) in which the first flower opens only 400 degree-days after initiation, which is the case of the fodder pea Winterberger.

    — in genetics: the main genes that control the transition from the vegetative to the reproductive stage have been identified and located on the genetic map of the pea;

    — in ecophysiology: dates of floral initiation and of the beginning of flowering have been modelled as a function of temperature and photoperiod, enabling efficient forecasting in varieties that are currently cultivated. These models are an indispensable tool for analysing the productivity of the crop.

In another connection, our knowledge of the physiology and molecular structure of the network of genes that control flowering in the model species *Arabidopsis thaliana* has considerably advanced. The information we have acquired thanks to gene sequencing of *Arabidopsis thaliana* and of the model legume *Medicago truncatula* already enables to identify homologous gene sequences responsible for flowering in pea. This new molecular tool is being used to accelerate the creation of plant material in which genetic variation is controlled and limited to a single gene, which will further enhance modelling. As these improved ecophysiological models are potentially more accurate, they in turn will help identify and rank factors that are still limiting in improving pea cultivation.

Reproductive development

Nathalie Munier-Jolain, Olivier Turc, Bertrand Ney

Associating the progress of developmental stages and environmental conditions helps us to understand the yield formation in a given situation and can help in the choice of appropriate practical techniques and genotype: characterization of the developmental stages of individual plant organs is a useful tool for analyzing the behaviour of genotypes in various environmental conditions and for predicting their performance; knowledge of the developmental calendar is thus a major component of models predicting plant growth and development.

For plants with a determinate growth habit, like maize or cereals, development of all the reproductive organs is closely synchronised. All the reproductive organs flower and achieve physiological maturity simultaneously, leading to a simple description of reproductive development. Conversely, for legume crops the sequential flowering up the stem leads to a wide heterogeneity of reproductive stages of organs which have not been fertilized simultaneously. Thus modelling reproductive development involves more than the prediction of the occurrence of a mean reproductive stage on the plant, and aims to predict the sequential development along the stem of flowers and seeds located at different morphological positions (mainstem or branches, node number, order of raceme).

Seed development

For all grain legume species seed development stages are similar: after fertilization seed development can be divided into three development phases:

- The first reproductive stage begins at fertilization and ends at "beginning of seed filling". During this period, cell division occurs in the embryos without significant dry matter accumulation (Ney *et al.*, 1993). At the end of this phase seeds are unlikely to abort (Duthion and Pigeaire, 1991) and the starch and protein reserves begin to accumulate within the cotyledons: this stage is called "beginning of seed filling" and corresponds to the final stage in seed abortion. The duration of this phase is variable among species: 300 (0°C base), 350 (0°C) base and 650 degree-days (3°C base) for pea, soybean, and white lupin respectively.

- The second phase begins at the stage "Beginning of seed filling" and ends at physiological maturity. During this phase cell division stops, near-linear dry matter accumulation begins in cotyledons and continues until physiological maturity (Ney *et al.*, 1993). At the end of this phase vascular connections between seed and mother plant are disrupted, corresponding to physiological maturity.
- After physiological maturity, assimilates are unlikely to be delivered to the seed and final seed weight is reached. The seed dries passively according to the air humidity until harvest.

Characterization of reproductive stages

- ### Which indicators?

The embryo is fertilised soon after the "open flower" stage, which constitutes a simple and reliable visual indicator of the beginning of seed development.

Duthion and Pigeaire (1991) have determined directly on green seeds the lengths corresponding to the final stage of seed abortion using a non-destructive photographic method to monitor seed lengthening. The lengths after which seed abortion does not occur are 8.5mm for pea, 12mm for soybean, and 13mm for white lupin. There is a close link between pod thickness and the length of the longest seed in the pod (Fougereux *et al.*, 1998). This relationship allows the pod thickness (which is easily measured) to be associated with the seed length specific to the final stage of seed abortion. For pea the final stage of seed abortion is reached when the pod thickness is 8.9mm (cv. Baccara). However the indictor based on seed length or pod thickness must be used cautiously: the critical sizes are very dependant on growing conditions. Moreover the seed length corresponding to the final stage in seed abortion probably depends on genotype.

The analysis of seed water concentration and seed dry matter confirms the close relationship between the moisture status of the seed and its growth and development. Hence seed water concentration is a convenient criterion to characterize the stages of seed development such as "Beginning of seed filling" and "Physiological maturity". As for seed growth and development, those stages established with water concentration fall into three phases:

- From fertilization, seed water concentration is stable at 85 % approximately;
- Seed water concentration decreases regularly until reaching a value specific to the species in question;
- Seed water concentration decreases rapidly and irregularly until harvest.

• Beginning of seed filling or final stage in seed abortion

The time when seed water concentration falls below 85 % is a useful indicator of the beginning of seed filling and of the final stage in seed abortion, which are concomitant.

For many legumes such as pea, soybean, lupin or chickpea, reserves start to accumulate within the seed when seed water concentration begins to fall below 85 %. At this stage various percentages of the final seed weight are already accumulated for pea, soybean and lupin, i.e. 10, 8 and 20 % respectively. This percentage is close to those represented by the seed coat weight in the total seed weight at maturity. Because seed coat expansion is almost ended when accumulation of reserves begins, seed weight at the beginning of seed filling corresponds approximately to the seed coat weight (Hedley *et al.*, 1994). Consequently the wide variability in seed weight between species at the beginning of seed filling is mainly due to the difference in seed coat weight.

The beginning of seed water concentration decrease coincides with the final stage of seed abortion: seed abortions occur exclusively when seed water concentration remains stable at 85 %. Whatever the nodal position of the pod within the plant and the number of seeds within the pod, the seed number counted when seed water concentration begins to fall is not significantly different from that at maturity (Munier-Jolain *et al.*, 1993; Dumoulin *et al.*, 1994).

• End of seed filling period or physiological maturity

The relationship between seed water concentration and relative seed growth (seed weight/ final seed weight) indicates a specific seed water concentration below which no more dry matter accumulation occurs in the seed. This threshold is variable among species, i.e. 55, 60, 60, and 65 % for pea (Dumoulin *et al.*, 1994), soybean (Munier-Jolain *et al.*, 1993), chick pea (Turc *et al.*, 1994) and lupin (Munier-Jolain *et al.*, 1997) respectively. For a given species the threshold is independent of genotype and environmental conditions. This stage corresponds to physiological maturity.

For all species seed development consists of three stages: fertilization, final stage of seed abortion which corresponds to the beginning of reserves accumulation within the seed, and physiological maturity which corresponds to the end of seed filling period. Seed water concentration is a convenient indicator of those stages. Seed water concentrations specific to each stages are stable among various environmental conditions, but vary between species.

Whole plant reproductive development

The three stages of seed development, namely flowering, beginning of seed filling, and physiological maturity, progress linearly along the stem as a function of cumulative degree-days (fig. 1.10). This pattern of whole plant reproductive development is similar in various environmental conditions (temperature and solar radiation) without stress, for pea (Ney and Turc, 1993) soybean (Munier-Jolain et al., 1993), and lupin (Munier-Jolain et al., 1997).

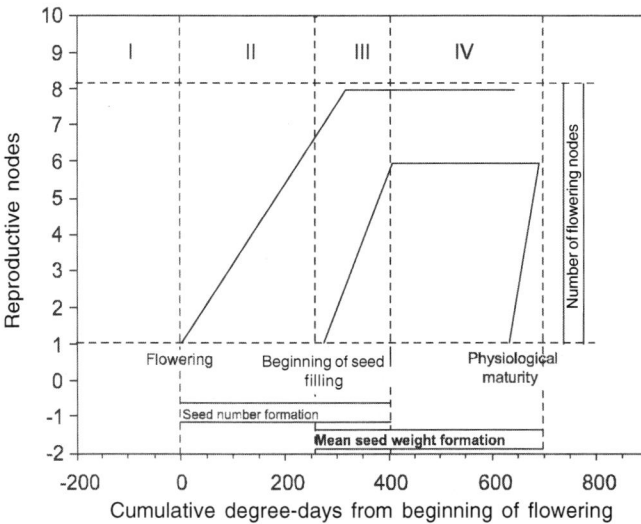

Figure 1.10. Diagram of the reproductive development model of a pea stem. This formalism allows to define 4 phases characterized or not by different sinks in competition for assimilates.

• Rates of progression of reproductive stages along the stem

The later the reproductive stage (from flowering to physiological maturity) the higher is the progression rate. In the same way the rate of leaf expansion is less than that of flowering (cf. fig. 1.3).

The rates of progression of each reproductive stage along the stem are constant over a wide range of environmental conditions (Ney and Turc, 1993, Munier-Jolain *et al.*, 1993) because the rate of progression of flowering is hardly sensitive to the plant growth rate (Ney and Turc, 1993). However when plant growth becomes very slow, the rate of progression of flowering along the stem may fall.

• Number of reproductive nodes

The indeterminate growth habit of grain legumes leads to a wide environmental and genotypic variability of the total number of reproductive nodes. For a given genotype the rank of the last reproductive node varies widely for various environmental conditions whereas the rank of the first reproductive node remains stable (cf. p. 19) (Roche *et al.*, 1998). Two hypotheses could explain the regulation of the total number of reproductive nodes: (i) hormonal signals produced by the already developed reproductive organs could limit the development of new reproductive organs; (ii) the pods bearing filling seeds could take priority for assimilate allocation and this competition for assimilates could prevent the production of new reproductive nodes.

Roche *et al.* (1998) proposed a model to simulate the total number of reproductive nodes in pea using three developmental parameters: (i) the initial node interval, corresponding to the number of developed nodes bearing leaves above the first reproductive node at the beginning of flowering; (ii) the rate of leaf expansion; and (iii) the rate of progression of flowering (fig. 1.11). The rates of leaf expansion and of progression of flowering are genotypic constants independent of environmental conditions (Turc and Lecour, 1997). Consequently the wide range in the total number of reproductive nodes is manly due to the initial node interval which depends both on genotype and environmental conditions (especially under limiting growth conditions).

• Duration of reproductive phases

The durations of each reproductive phase, i.e. flowering-beginning of seed filling and beginning of seed filling-physiological maturity, decrease with the rank of the node, because of the increase in the progression rate of the later reproductive stages.

The duration between flowering and beginning of seed filling for the first reproductive node remains relatively stable at 270 degree-days among various environments without any biotic or abiotic stresses. Conversely the duration of seed filling (from beginning of seed filling to physiological maturity) is highly variable for various environmental conditions even in

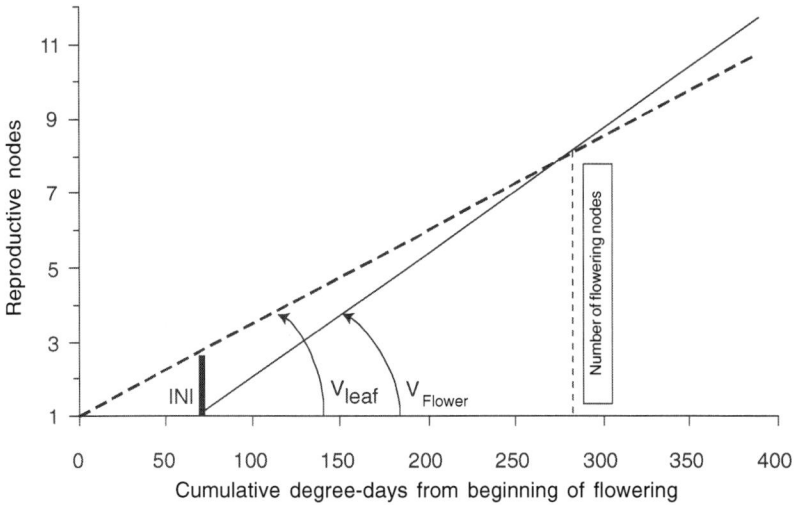

Figure 1.11. Rate of progression of leaf expansion (--) and of flowering (—) along the stem. The 3 parameters INI (Initial Mode Interval) V_{leaf} (rate of leaf expansion) and V_{flower} (rate of progression of flowering) allow to simulate the total number of reproductive nodes (Roche *et al.*, 1988, reprint with the kind authorization of Oxford University Press, Ann. Bot. 1998, 81(4), 545–555).

optimal growth conditions, depending on the source-sink ratio at the end of the growth cycle (see p. 114–115). However both duration between flowering and seed filling period and duration of seed filling decrease under strong water or nitrogen deficits (Sagan at al., 1993; Ney *et al.*, 1994).

The total number of reproductive nodes varies greatly, especially due to variation in environmental conditions (Roche *et al.*, 1998) or to the genotype (Dumoulin *et al.*, 1994). For a given genotype this number can double when plant density is very low without any modification in the rate of progression of flowering. This huge variability in the total reproductive nodes number causes major variations in the lengths of the reproductive phases at the plant scale.

How to represent whole plant development diagrammatically?

Knowledge of reproductive stages associated with the model simulating the beginning of flowering allows reproductive development along the stem to be modelled. In conventional agricultural conditions all the stems in the canopy contribute similarly to the crop yield, the main stem and branches having similar reproductive development. This model could be used:

- *To predict*: a small set of measurements gives a precise description of the reproductive development from flowering to harvest. The beginning of flowering at the first reproductive node could be predicted using solely weather data. In optimal environmental conditions without any biotic or abiotic stresses, the beginning of seed filling occurs approximately 270 degree-days after the beginning of flowering. The stage "Beginning of seed filling" progresses linearly up the stem at a rate which is constant for each genotype. Consequently the measurement of the total number of reproductive nodes and the occurrence of physiological maturity at one reproductive node can establish the time course of reproductive development on the whole plant scale;

- *To evaluate*: the diagrammatic representation of reproductive development on the whole plant scale can provide a precise estimate in a given environmental condition of the periods of formation of yield components, seed number and individual seed mass, on a plant or even a node basis. The identification of the periods of formation of yield components is useful to relate the yield components to the environmental conditions occurring during their formation for a given node;

- *To understand*: a knowledge of the general pattern of reproductive development allows the reproductive stage, the morphological location on the stem, and the number of all the reproductive organs on a stem to be identified at any time. This tool is useful to understand and model the physiological processes of seed production, especially when they are linked to competition for carbon or nitrogen assimilates. Figure 1.10 shows the different phases characterized by the existence or otherwise of competitive sinks for assimilates. During the period preceding flowering, vegetative sinks (shoot, roots, nodules) compete together for assimilates (phase I). From the beginning of flowering at the first reproductive node until the beginning of seed filling at this node, young pods bearing seeds with little dry matter accumulation represent new sinks for assimilates (phase II). Phase III is the most complex for analyzing assimilate competition, because two types of reproductive sink compete together with vegetative growth: the young pods at the top of the plant and the older seeds on the lower reproductive nodes of the stem which begin to accumulate reserves in their cotyledons. This period is a characteristic of the indeterminate habit of pea. The duration of the phase is highly variable depending on the total number of reproductive nodes. During phase IV, which begins when the youngest pods at the top of the plant reach the beginning of seed filling, filling seeds represent the main sink for assimilates at the expense of vegetative growth and nitrogen fixation.

Branch emergence

Christian Jeudy, Nathalie Munier-Jolain

Characteristics of pea architecture

The initiation and phenology of lateral branches in pea have been less studied than for the main stem. However lateral branches often contribute significantly to the yield, but with a genotypic variability. For winter peas, chilling of the main stem apex increases the proportion of branches in the canopy for ensuring the growth in spring.

For peas, all the axes are constructed from successive phytomers (see p. 5). In the axil of a leaf a phytomer can bear an axis which is at first a lateral bud likely to produce a branch of a higher order:

- When a bud, the simplest form is a meristem. The differentiation and thus the initiation within the bud produce more or less mature phytomers (Aitken, 1978).
- If the bud becomes a branch, ie a stem, the initiated axis appears in the plant phenology.

The main stem corresponds to the first axis of the plant (1st order). The lateral branches, whether developed or not, located in the axil of a leaf on the main stem constitute the 2nd order of axis, and so on for further orders of axis.

In order to identify an axis on the plant, the terminology proposed takes into account the position of the considered axis on the lower axis which bears it (fig. 1.12).

Ri identifies a lateral axis borne by node i of the main stem. When many buds are located at the same node, the second bud is called "prim" (p) and the next "second" (s) and so on.

Ri-j identifies a lateral axis borne by the node j on the branch Ri.

For spring pea the 2nd order axes generally make up most of the branches. They are located at the bottom of the stem, especially on the 2nd node (axis R2 according to the proposed terminology).

Note: the numeric data proposed below were essentially obtained for the cultivar Baccara. These data are proposed as an example but cannot be generalized because of the huge genotypic and environmental variability (architectural type and growth conditions).

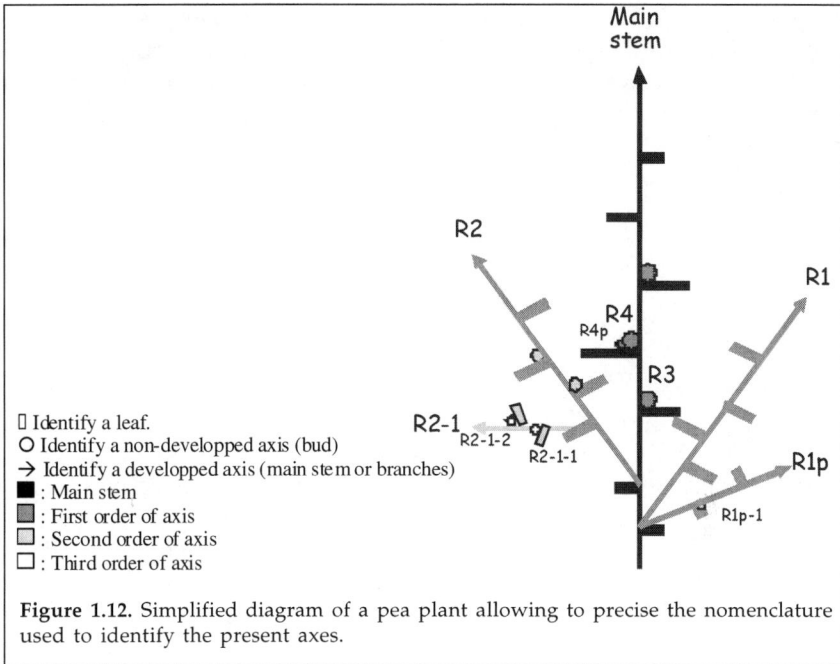

Figure 1.12. Simplified diagram of a pea plant allowing to precise the nomenclature used to identify the present axes.

Initiation of meristems

• The number of meristems, a genotypic characteristic

The lateral axes are generally observed on vegetative nodes (Arumingtyas, et al., 1992) and are rarely present on reproductive nodes. The frequency of lateral axes along the vegetative nodes exhibits genotypic variability (Jeudy, 2001). The number of lateral buds on the main stem is maximal for the nodes located at the base of the stem (up to 4 buds per node) and decreases to one (sometimes 0 according to the genotype) for the upper vegetative nodes. The decrease in bud number from the base to the top of the stem may be due to the apical dormancy which inhibits bud initiation.

For a given genotype the number and the position of lateral buds are barely affected by the growing conditions (Pigeaire, 1986) and thus constitute a genotypic characteristic.

• Meristem initiation linked to the initiation of phytomers

The initiation of lateral meristems happens early in the plant growth cycle. At emergence, 8 to 9 lateral buds are already initiated on the first five vegetative nodes of the main stem. On the first and second nodes, the axillary meristems "prim" and "second" are also already initiated. The buds are initiated from the base to the top of the axis on which they appear.

For a given axis two different activities of organogenesis occurs concomitantly: on the one hand new phytomers are initiated in the apical meristem, and on the other, lateral meristems appear in the axils of previously initiated phytomers. The duration between the initiation of two successive lateral meristems is approximately equal to the plastochrone of the axis on which they are located. Consequently the delay between the initiation of a new phytomer and the initiation at its axil of a lateral meristem is constant.

• Initiation of phytomers

At emergence several phytomers are already initiated in each bud. The older the lateral axis, the greater is their number.

The apical meristems of branches (i.e. developed lateral axes) exhibit initiation activity during all their growth period, unlike latent axes (buds) whose activity decreases progressively with the initiation of only few phytomers. For a given order of branches, the rates of phytomer initiation are constant when expressed in degree-days (0°C basis). Conversely plastochrons vary significantly between orders of axes. The rate of phytomer initiation by apical meristems is slightly higher on branches than on the main stem. For example for the cultivar Baccara, the plastochrons last 46 and 38 degree-days for the main stem and R2 branches respectively (Jeudy, 2001).

• Floral Initiation

As for the main stem, lateral buds are sensitive to environmental conditions, mainly photoperiod and temperature. Thus whatever the axis, the level of the first reproductive node is higher for plants grown in a short day photoperiod than in long days.

Floral initiation happens synchronously for the main stem and the first initiated axes (such as R2 branches), approximately 170 degree-days after emergence (Jeudy, 2001). Consequently in spring pea crops whose canopy is mainly made up of main stems and R2 branches, flowering is synchronous for all the stems. For winter peas, floral initiation on branches located on scale leaves is synchronous with the main stem (Lejeune, Pers Comm).

For the other upper lateral axes, the later the axis appears, the later its floral initiation occurs. However the later the axis appears, the shorter is the delay between its appearance and its floral initiation. Consequently the first reproductive node is located lower on the stem, with a smaller number of vegetative nodes than on the main stem.

• Modelling the potential architecture

For spring peas in a given environment, knowing the rules of appearance of lateral axes (position and rate of initiation), their plastochron, and their timing of floral initiation, allows the chronology of the main stages linked with organogenetic activities to be modelled in degree-days, and thus the potential architecture of the plant to be established: each axis is identified by its position, number and nature of initiated phytomers. However the model must be parameterized for each genotype and for the growing conditions (especially photoperiod).

Growth of branches

Only some of the initiated organs which form the potential architecture grow up to be part of the plant phenology.

In optimal growing conditions the chronology of the appearance of lateral axes is linked to that of the lateral buds, from the base to the top of the plant. For spring peas the upper nodes rarely bear branches.

• The position of branches

The position of branches seems to depend on the apical dominance (Cline, 1994) whose setting in and removal is under hormonal control (Cline, 1991; Nagao & Rubienstein, 1976). Auxin synthesized by the apex and the youngest leaves diffuses from the top to the bottom of the plant and acts as an inhibitor to the growth of lateral buds. Conversely cytokinins produced by the stem and the roots are diffused from the bottom to the top of the plant and promote the growth of lateral buds (Cline, 1991). Consequently the gradients of auxin and cytokinins are opposite within the plant and thus the ratio "auxin/cytokinins" at a given node is well correlated with the growth of the axillary buds at this node (Emery et al., 1998; Cutter & Chiu, 1975). As a consequence, in many agricultural conditions, branches are located only on the lowest vegetative nodes (Husain & Linck, 1966; Doré, 1994) where the ratio "auxin/cytokinins" is the most favourable to the growth of branches. The apical dominance is high at the beginning of the growth cycle, but decreases progressively until the appearance of reproductive stages; thus branches should appear later on the upper nodes.

The growth of branches in apical positions can also be explained by the modification of light quantity and quality received by each bud during canopy closure. The light quality, and especially the red/far red ratio, influences the morphogenesis of branches (Cutter & Chiu, 1975): applying far red increases apical dominance (Cline, 1991, Healy et al., 1980). Thus at high density where plants are shaded, the light quantity declines at the bottom of the plant with a concomitant increase in the proportion of far red. The growth of lateral buds is then inhibited (Casals *et al.*, 1986).

> Environmental conditions (light, hormones) affecting each bud locally are likely to determine its ability to develop. In spite of this ability, the effective growth of the branches happens only if enough assimilates are available within the plant.

• The number of branches

Although the developed axes appear generally from the bottom to the top of the plant, the determination of the sequence of appearance of grown branches remains unknown. Moreover the appearance of R1 is highly variable from one situation to another.

The use of thermal time and the developmental stages on the main stem does not allow the timing of the appearance of branches to be predicted. The number of branches for each order seems to depend on the total plant growth (Jeudy, 2001). If mineral or water nutrition are limiting, branches do not start to grow (Cline, 1991; Cutter & Chiu, 1975; Doré, 1994; Nougarède & Rondet, 1978; Bakry *et al.*, 1984; Rubinstein & Nagao, 1976). The various levels of competition for assimilates, as affected by plant density, affect the branch growth by modifying the total plant growth rate (Pigeaire, 1986; Casal *et al.*, 1986; Bakry *et al.*, 1984; Knott & Belcher, 1998). Thus the lower the plant density, the more numerous are the orders of branches (fig. 1.13).

• The type of branches

In a canopy with a high plant density, differences in sowing date have no effect on the type and frequency of the branches which grow. For the more sparse plant densities the number of branches varies widely with the sowing date, without variation in dry weight: the earlier the sowing date, the larger is the number of branches. This could be explained by (i) the positive effect of low temperatures on branch growth (Doré, 1994; Stanfield *et al.*, 1966) and (ii) to a lesser extent, the positive effect of a short photoperiod on the initiation of branch growth (Field & Jackson, 1974). However genotypes react differently to photoperiod (Arumingtyas *et al.*, 1992; Paton, 1968). Whatever the sowing date and the plant density, R2 branches are always present in the canopy.

• R1 is an atypical branch

For the R1 axes, the plastochron, floral initiation and growth or otherwise of the bud into a branch are highly variable in different growing conditions, and even between plants within the canopy. This variation is due to the position of this lateral axis in relation to the air/soil interface. Depending

Mean number per plant

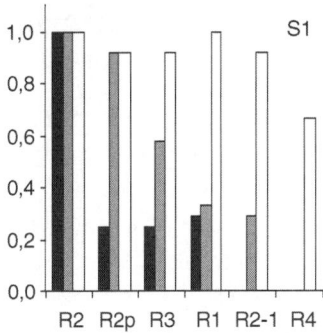

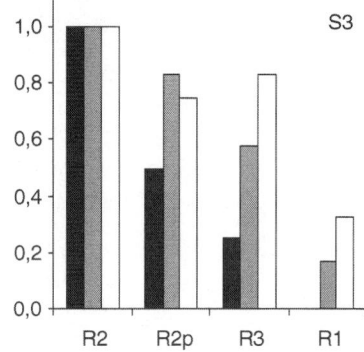

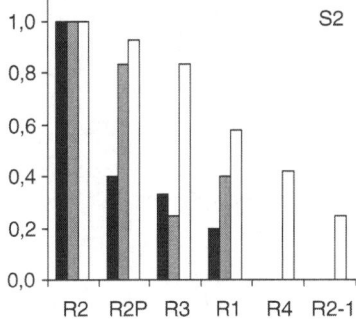

Mean number per plant

Mean number per plant

Figure 1.13. Distribution of different types of axes for three spring sowings: S1 (early sowing), S2 and S3 (late sowings). (□=10 plants.m^{-2}; ▦=40 plants.m^{-2}; ■=80 plants.m^{-2}). The number of axes showed in the figure corresponds to the maximal frequency observed for the corresponding axis.

on the sowing depth (Doré, 1994) or the soil surface structure, the amount of light received by the R1 axis varies between plants, leading to a non-uniform activity of organogenesis or morphogenesis within the canopy.

• Phyllochron of branches

For the commonly used spring pea cultivars, phyllochrons of branches are similar to that observed on the main stem. However in extremely unfavourable growing conditions or for very low plant densities, there is a wide range of growth rates on the various axes and thus an inverse relationship between phyllochron and growth rate of a given axis: in poor growing conditions an increase in the phyllochron is linked to the reduced growth rate of the axis.

• Flowering

Field grown canopies are generally made up mainly of basal branches. When growing conditions are favourable (free of stress) all the stems (main stems and branches) flower simultaneously.

When many orders of axes are grown, for instance at low density or when upper lateral axes grow after a period of stress, the lateral axes should flower later than the main stem. This delay depends on:

- The time when the branch in question begins to grow; this date depends on assimilate availability in the plant;
- The number of vegetative nodes on the lateral axis, which depends both on the timing of floral initiation and the position of the considered axis;
- The phyllochron of the considered axis, which depends on its growth rate.

Conclusion

The pea plant has a substantial architectural potential. Unlike white lupin (Munier-Jolain, 1996) the number of buds is unlikely to limit the number of branches. In spite of genotypic differences, there are usually enough buds to allow the development of a large number of branches, whatever the growing conditions. This characteristic enables the pea crop to regrow with new stems after cultural (pest damage to the upper nodes) or climatic (apex chilling) accidents.

> In practice in field grown peas the basal branches (R2, branches on the 2nd scale leaf of the main stem) make up the majority of the axes. Consequently in most cases with normal plant density and avoiding major stresses, the stems (main stem or branches) are slightly different; thus the responses of main stem and branches to environmental conditions can be considered as similar.

Methodological sheet: measurements of developmental stages
Isabelle Chaillet, Véronique Biarnès

Measurements of stages at the stem scale

- Emergence

The emergence stage is reached when the second scale is visible.

- Foliar stage: from emergence to the "Beginning of flowering" stage

The foliar stage of a plant corresponds to the number of developed leaves on the main stem.

The two first nodes do not bear real leaves, but the two scales. They are not to be taken into account when the number of leaves is counted. Those scales bear generally the branches.

A leaf is taken into account for the calculation of the number of leaves when it is completely expanded (stipules completely or fully opened). For instance, in figure 1.14, the stem is at the stage 7 leaves (7 expanded leaves). The eighth leaf is in development: stipules are still folded one against the other. For the winter pea, the main stem is often aborted and the number of leaves is evaluated on the more developed branch (fig. 1.15).

- Floral initiation

The floral primordium appears at the level of the cauline meristem (Cf fig. 1.6). It is observed on the apex with a binocular magnifying glass, after removing the enclosing leaves.

The floral initiation stage is then difficult to observe. It is thus possible to locate it from a foliar stage: 3 to 4 leaves for spring pea and 5 to 6 leaves for winter pea.

- Beginning of flowering

Beginning of flowering: the 1st flower is well opened (stage 0,5 in the scale of Maurer *et al.*, 1966; fig. 1.16).

In a field, for spring pea, the beginning of flowering stage occurs 2–3 days after the appearance of the first flowers in the crop.

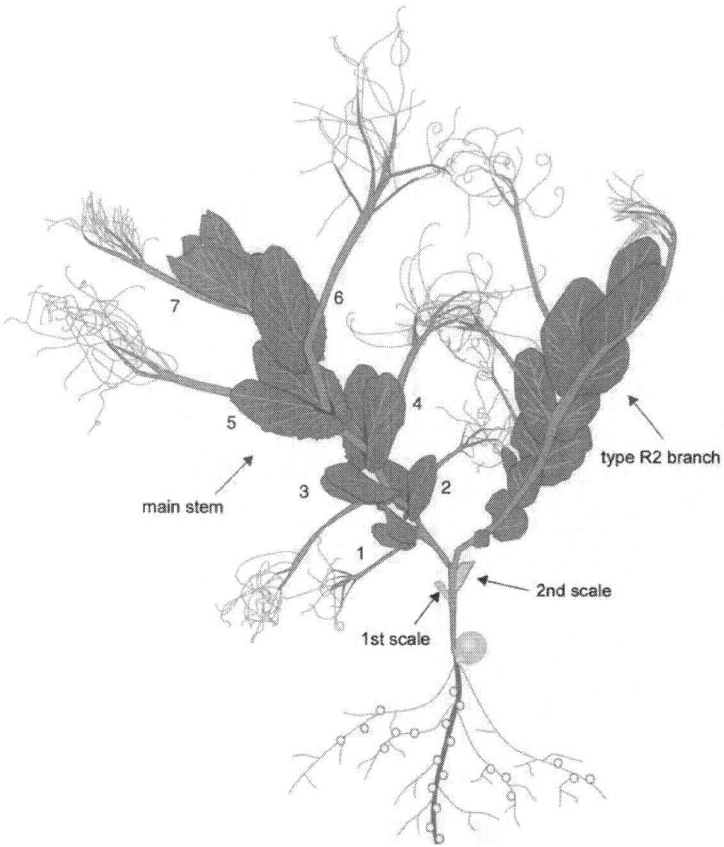

Figure 1.14. Example of a pea plant at 7 leave stage. This plant has a main stem and one branch. The main stem has 7 leaves, numbered from 1 to 7.

• Flowering stage: from beginning to end of flowering

Flowering stage: number of flowering nodes or flowering of one node level.

The last flower to be taken into account for the calculation is the last one well opened at the top of the plant (fig. 1.16). In case of early abortion of pods or of non-fertilization, a node is considered as a flowering node when a floral peduncle of at least 1 cm long can be observed.

In the example presented in figure 1.17, stage is flowering at the 7th node. Pods (at the lower nodes) and flowers (at the 3 upper nodes) are simultaneously present on the plant.

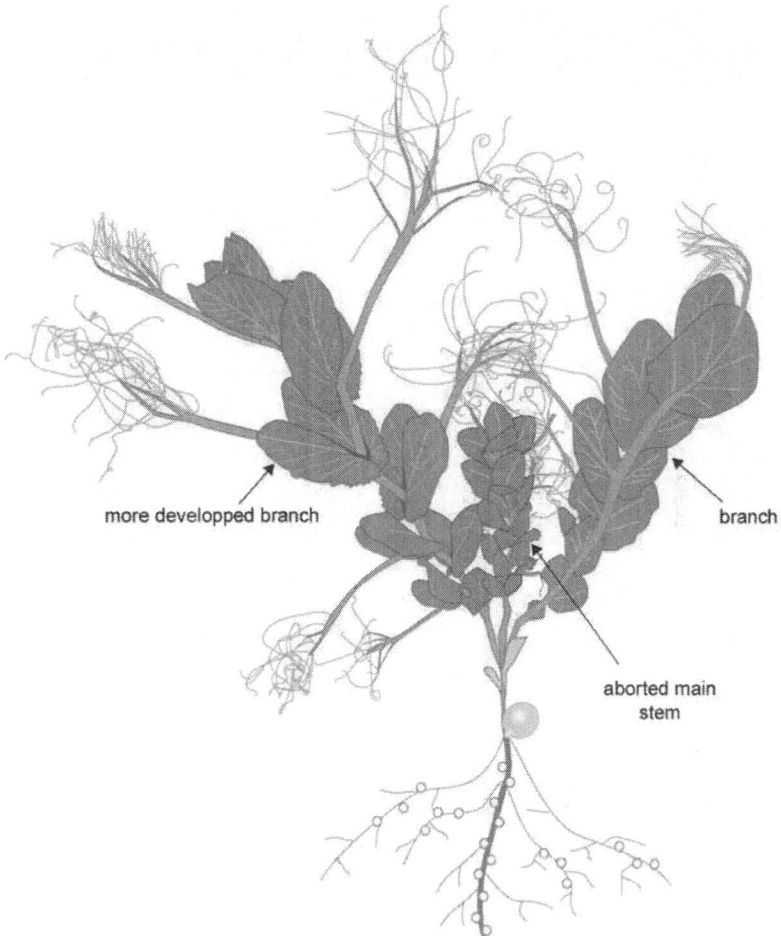

Figure 1.15. Example of a winter pea plant, with stopped main stem and two branches. In this case, for the notation of foliar stage, we can count the number of leaves on the more developed branch.

- ## Reproductive stages: from beginning to end of final stage in seed abortion

Reproductive stage: number of the last node bearing at least one pod that has gone through the final stage in seed abortion (FSSA). We note the number of the last node and not a number of reproductive nodes because it can be nodes without pods and it must be taken into account for the measurement of the stage.

At a given node, stages FSSA (final stage in seed abortion) and BSF (beginning of seed filling) are concomitants (Cf. p. 25).

Figure 1.16. One opened pea flower— Stage 0,5 in the scale of Maurer *et al.* (1966) (Source UNIP).

Figure 1.17. Stage: flowering at the 7th node. The stem has 7 flowering nodes or having flowered: nodes numbered from 1 to 7.

A pod is at FSSA when at least one of its seeds has gone through 8 mm long, that corresponds to a thickness of the upper pod over or equal to 7 mm. For this measurement, Arvalis-Institut du Végétal has developed a tool to calibrate pods, that allows knowing quickly if a pod has reached or not the FSSA stage. An other method, heavier but more accurate, consists to follow the evolution of seed water concentration, the stage FSSA is reached when seeds water concentration of a pod decreases under 85%.

. When the last reproductive node has gone through FSSA, the number of seeds is determined (Cf. fig. 1.10).

• Physiological maturity

Physiological maturity stage is reached when the number of the last node from one pod at least bears seeds with water concentration less than 55 %.

At physiological maturity, seed filling is finished: yield is determined.

Measurements of stages at the crop scale

The crop has reached a given stage when a percentage of stems, variable depending on the stage, has reached these given stage.

The number of stems or area mentioned in the table annexed, corresponds to the total number to be picked up for each measurement. In trials with replicates, measurements must be made on several replicates.

For the determination of the date of beginning of flowering in a plot, it is possible to visit the plot only one time, when plot is fully flowered. The flowering stage is measured with the mean number of flowered nodes, the date of beginning of flowering can be estimated taking into account a delay of 40 degree-days for the flowering of two successive nodes. For example, if the plant has 3 flowering nodes, 80 degree-days must be removed from the date of notation to estimate the date of beginning of flowering.

Stage	Size of samples*	Number of samples	Overstepping of a stage	Method
Emergence	8 x 1m² (2 consecutive rows)	4 from emergence to stage 2–3 leaves	80 % of the plants have gone through the stage	Counted in field
Foliar stage	4 x 15 consecutive plants			Destructive foliar notation
Floral initiation			50 % of the plants have floral apices	Observation with a binocular magnifying glass destructive
Beginning of flowering	4 x 15 consecutive stems	1 to 3	50 % of the plants have gone through the stage	In the field or destructive measurement
Flowering stage	4 x 15 consecutive stems			Destructive notation
Reproductive stage	4 x 15 consecutive stems			Calibration of pod or measurement of the seed water concentration(48 h to 105°C in oven)
Physiological maturity	4 x 10 to 15 consecutive stems		50 % of the plot area has turned to yellow	Measurement of the seed water concentration (48 h to 105°C in oven)

Indicative time position of pea developmental stages

Stages	Mean duration in cumulative degree-days*
Emergence	150 degree-days for spring sowings. 200 degree-days for autumn sowings.
Foliar development	Leaf appearance rate : 50 to 55 degree-days between two foliar nodes for most of the spring varieties. Some varieties exhibit weaker values: from 40 to 45 degree-days.
Floral initiation	150 to 200 degree-days from emergence for spring sowings.
Beginning of flowering	675 to 750 degree-days from emergence to the date of beginning of flowering for spring pea varieties.
Floral development	40 to 45 degree-days between two flowering nodes for pea spring varieties.
Final stage in seed abortion (FSSA)	25 to 30 degree-days between two reproductive nodes for both winter and spring pea varieties. Around 270 degree-days from flowering to FSSA of the first reproductive node for spring pea varieties.
Physiological Maturity (PM)	20 to 25 degree-days between two successive nodes. 300 to 500 degree-days from BSF to ESF for a given node. The duration of the desiccation phase from stage PM (water concentration: 55 %) to harvest stage (water concentration: 14 %) is highly variable depending on the temperature and on the air hygrometry, from 5 to 15 days or more.

BSF: Beginning of seed filling (concomitant with FSSA)
ESF: end of seed filling
*   The durations expressed here correspond to cumulative degree-days (temperature base 0°C). For foliar development, floral development, final stage in seed abortion and physiological maturity those durations represent respectively the inverse of rate of appearance of leaves, of flowering progression, of FSSA progression and of physiological maturity progression along the nodes. Those rates of progression may be modulate according to genetic and environmental variability (cf. part I.1).

2

Carbon acquisition at the crop level in pea

Lydie Guilioni and Jérémie Lecoeur

Introduction

Numerous studies since the 1960s have shown a quantitative relationship between intercepted radiation and growth (see review by Sinclair and Muchow, 1999). Monteith (1972, 1977) first made the distinction between the crop function in intercepting solar energy and its function in transforming intercepted energy into biomass. He proposed a theoretical approach linking biomass production to incoming radiation through biophysical conversion factors (Eq.1).

$$DM = \int_{t1}^{t2} RUE \ RIE \ CE \ R_s \ dt \qquad \text{(Eq. 1)}$$

with DM dry matter produced between the times $t1$ and $t2$, Rs incident solar radiation, RUE radiation use efficiency , RIE radiation interception efficiency, CE climatic efficiency.

Monteith's formula presents crops as structures able to intercept solar radiation (Rs) and to convert it into biomass, with particular efficiencies in each step (climatic, radiation interception and radiation use efficiencies). This approach has become a robust and often-used method for analysis and modelling of crop growth. In conditions where neither water nor mineral nutrient are limiting and in the absence of pests and diseases, solar radiation is the main physical limit to crop yield. Given an incident radiation, the productivity of a crop depends on its ability to intercept and then use the radiative energy.

This modelling approach of biomass production is used hereafter to analyse the carbon assimilation of pea at the crop level. This approach requires the estimation of the different efficiencies and the comprehension of their sources of variation. Parts 1 and 2 define the different terms of Monteith's equation and describe some ways of estimating them. Comparing results from the literature is often difficult because of changing and sometimes approximate terminology. This difficulty can also be due to various and not clearly defined methods used for estimating the terms of Monteith's formula. Parts 3 and 4 present the sources causing variation in radiation interception and radiation use efficiencies for a pea crop.

Definition of the terms of Monteith's equation

Monteith's formalism is based on the radiative balance of the crop. Nevertheless, depending on the chosen radiation, equation 1 can present different forms. Total solar radiation (wavelength from 0.3 to 3 μm) or the photosynthetically active radiation (PAR, wavelength from 0.4 to 0.7 μm) can be used. Some authors estimate intercepted radiation while others use absorbed radiation. These slight differences have consequences on the definition of the various efficiencies. It is necessary to understand this to be able to compare the different results from the literature.

• Radiative balance of a canopy

Solar radiation is a shortwave radiation (0.3–3 μm) composed of direct and diffuse components. Direct beam solar radiation comes in a direct line from the sun, depending on the sun's position defined by solar elevation and azimuth. The diffuse solar radiation component is scattered out of the direct beam by molecules, aerosols, and clouds of the atmosphere. It comes from all parts of the sky. The wavelengths of solar radiation involved in biomass production are the photosynthetically active radiation between 0.4 and 0.7 μm. The PAR proportion to total solar radiation, so called climatic efficiency (CE), is almost constant in natural conditions and equal to 0.48. This allows equation 2 to be written :

$$DM = \int_{t1}^{t2} RUE \ RIE \ PA \ R_0 dt \qquad \text{(Eq. 2)}$$

with DM dry matter produced between the times t1 and t2, PAR_o incident photosynthetically active radiation, RUE radiation use efficiency, RIE radiation interception efficiency.

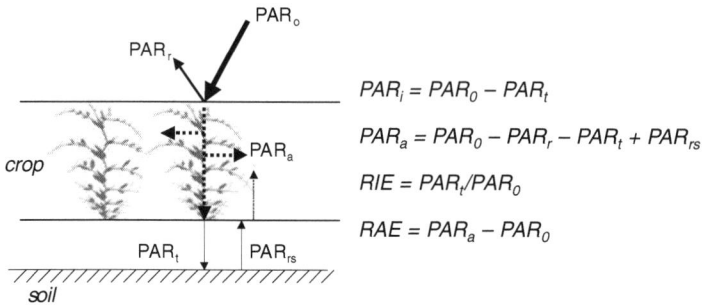

$$PAR_i = PAR_0 - PAR_t$$

$$PAR_a = PAR_0 - PAR_r - PAR_t + PAR_{rs}$$

$$RIE = PAR_i/PAR_0$$

$$RAE = PAR_a - PAR_0$$

Figure 1.18. Interactions of photosynthetically active radiation with a crop and related definitions (see Varlet Grancher *et al.*, 1989). PAR_0: incident photosynthetically active radiation, PAR_a: photosynthetically active radiation absorbed by canopy, PAR_i: photosynthetically active radiation intercepted by canopy, PAR_r: photosynthetically active radiation reflected by canopy, PAR_t: photosynthetically active radiation transmitted to soil, PAR_{rs}: photosynthetically active radiation reflected by soil, RIE: interception efficiency of incident PAR, RAE: absorption efficiency of incident PAR.

Radiation interaction with a plant canopy is complex. Incident radiation is reflected, absorbed or transmitted (fig. 1.18) with differing proportions according to the considered wavelength. The amount of intercepted radiation is defined as the difference in the downward flux density of radiation above and below the canopy. The amount of absorbed radiation represents the radiative budget of the crop that is the balance between the sources of radiation for the canopy (incident and scattered radiation) and losses of radiation (reflected radiation from both soil and vegetation, radiation transmitted to the soil). Radiation interception efficiency (RIE) is calculated by dividing the amount of intercepted radiation by the amount of incident radiation. The radiation absorption efficiency (RAE) is the ratio of the amount of absorbed radiation to the amount of incident radiation. Figure 1 illustrates these definitions for the visible wavebands (PAR). A detailed review of these concepts is proposed by Varlet-Grancher *et al.* (1989).

• Radiation use efficiency RUE

The radiation use efficiency of a crop can be defined as the ratio of the biomass produced to the amount of captured radiation. This general definition can be used in numerous ways. Most of the time only aerial biomass is taken into account but total crop biomass can be used. Captured radiation can be expressed as either total solar radiation or that in the wavebands of PAR. Finally, captured radiation can refer to either intercepted or absorbed radiation. Consequences of these choices on the determination of RUE are analysed by Bonhomme (2000). In the part 4 of this article, RUE is defined as the ratio to the aerial biomass to the amount of intercepted PAR:

$$RUE = \frac{\Delta DM}{\Delta PAR_i}$$

(Eq. 3)

Methods of estimation of the terms of Monteith's equation

The objective is not to draw up an exhaustive list of the existing ways to estimate the different terms of equation 2, but rather to report the most commonly used methods in crop physiology studies, especially for pea and to present some limits of these methods. Varlet Grancher et al. (1989), Sinclair and Muchow (1999) discussed these aspects in greater depth.

Biomass determination

The crop mass used in the calculation of RUE is usually based on net above-soil biomass production because roots are difficult to assess. Including roots will result in higher RUE, especially in the early plant cycle. The main source of errors in estimating biomass comes from crop heterogeneity and sample variability. Even when the crop is relatively homogeneous (density, plant size), biomass estimation is accurate to only 5–15 % (Gallagher and Biscoe, 1978). In order to limit the amount of skew in estimating biomass produced between two dates, it is thus preferable to take several samples on each date in different zones of the parcel. The samples must be sufficiently large and distant from each other to avoid the effects of overlapping. Indeed, a lateral input of radiation would lead to an over-estimation of RUE. On farmers' parcels, an acceptable minimum sampling would consist of four samples of 1m² per date. The samples are weighed after 48 hours drying at 80°C.

Assessment of radiation absorption or interception efficiencies

- By means of measurements of the different components of the solar radiation balance

All the terms of the solar radiation balance can be measured by using radiation sensors placed in four locations on the parcel:

— a horizontal sensor above the crop facing the sky measuring incident radiation on the canopy,

— a horizontal sensor above the crop facing the soil measuring radiation reflected by the canopy,

— a linear sensor or several specific sensors placed horizontally to soil level measuring incident radiation on the soil or transmitted by the canopy,

— a horizontal sensor facing the soil at the level of the first leaves measuring radiation reflected by the soil.

In this way, intercepted or absorbed radiation can be calculated and corresponding efficiencies (RIE or RAE) can be estimated. The main difficulty lies in accounting for the spatial heterogeneity of radiation transmitted on the soil. Canopy structure heterogeneity needs to be taken into account, especially at the beginning of the cycle when the inter-row is still visible. According to Muchow *et al.* (1994), quoted by Sinclair and Muchow, (1999) four line quantum sensors are required to account for canopy heterogeneity over a well-managed and relatively homogeneous agricultural parcel.

Readings can be taken on an on-going basis throughout the experimentation period but they tend to tie up equipment, which is both expensive and specialised (sensors and data acquisition). Techniques have been developed for taking intermittent readings. The limitation of these methods lies mainly in extrapolating an intermittent reading over a day. The fraction of radiation intercepted by a canopy depends on the angle of incidence of the sun's rays (i.e. time of reading) and on weather conditions, particularly the proportion of direct or diffuse radiation. Hence, readings should be taken at a regular fixed time and overcast periods should be avoided on sunny days. The second caveat relates to the interpolation between different reading dates. It leads to an error in the estimation of daily RIE in inverse proportion to the frequency of readings. The problem is not peculiar to this method and relates to all cases of intermittent readings.

• By means of Beer's law and the leaf area index

Radiation interception by a vegetal canopy varies according to its leaf area index (LAI). A relatively simple analysis can be carried out by transposing to the canopy a law on the attenuation of a monochromactic radiation passing through a homogeneous medium, known as Beer's law.

According to Beer's law, illumination reaching z depth (PAR(z)) is expressed:

$$PAR(z) = PAR_o \, e^{-k \, LAI_{(z)}} \qquad \text{(Eq. 4)}$$

with PAR_o illumination at the top of the crop canopy, $LAI_{(z)}$, leaf area index at z depth cumulated from the canopy top and k radiation extinction coefficient.

In this way, radiation reaching the soil can be calculated and radiation interception efficiency can be deduced from it:

$$RIE = 1 - e^{-k\,LAI} \qquad\qquad (Eq.\,5)$$

Using Beer's law to estimate RIE assumes that the vegetal canopy acts as a totally homogeneous diffusing medium. For this, leaves should be distributed randomly in the area occupied by the vegetal canopy. However, pea is a row crop. This canopy structure can be visible until the beginning of flowering. During this first phase of the cycle, the canopy consists of heterogeneous zones with high foliage density on the row and few or virtually no leaves on the inter-row. With a concentration of foliage in certain zones of the area, the canopy is less effective at intercepting radiation than if the foliage was distributed homogeneously. The result is an over-estimation of RIE with Beer's law at the beginning of the cycle (Lecoeur and Ney, 2003, fig. 1.19a). For LAI greater than 2, the use of Beer's law gives a good estimation of RIE. For lower LAI values, it leads to an over-estimation of on average 20 %. The longer the phase where LAI is lower than 2, the greater this over-estimation will be. For spring pea, this over-estimation does not have a significant impact on yield estimations because it relates to a period when biomass production is weak. For winter pea, on the other hand, the period with a weak LAI is long and this estimation can result in errors on the estimation of the canopy state at winter's end. This over-estimation can be corrected by taking into account soil cover rate by the vegetative canopy. For that, it is sufficient to consider separately the zones covered by vegetation and the zones of bare soil by calculating RIE separately only for the zones covered by vegetation. This approximation has the advantage of being easily applicable since cover rate can be estimated by means of a photograph or even by the naked eye for a trained practitioner.

– LAI estimation

LAI can be measured using several methods. In all cases, representative sampling of the plants in the particular zone needs to be carried out. Leaf area can be calculated by actually measuring the area of each leaf with a planimeter. Leaf area can also be calculated by using allometric relationships between leaf length (stipule length, in particular) and area. Finally, leaf area can be estimated on the basis of its dry biomass. This entails determining first of all a conversion coefficient of the dry foliar matter on area or area mass. These approaches require no or very little specialised equipment, except for a planimeter. On the other hand, they do require LAI to be measured several times during the cycle and are therefore very time-consuming. Approaches combining measurements and modelling also make it possible to reconstruct the course of LAI *a posteriori*. Their principle is to estimate the leaf area of each

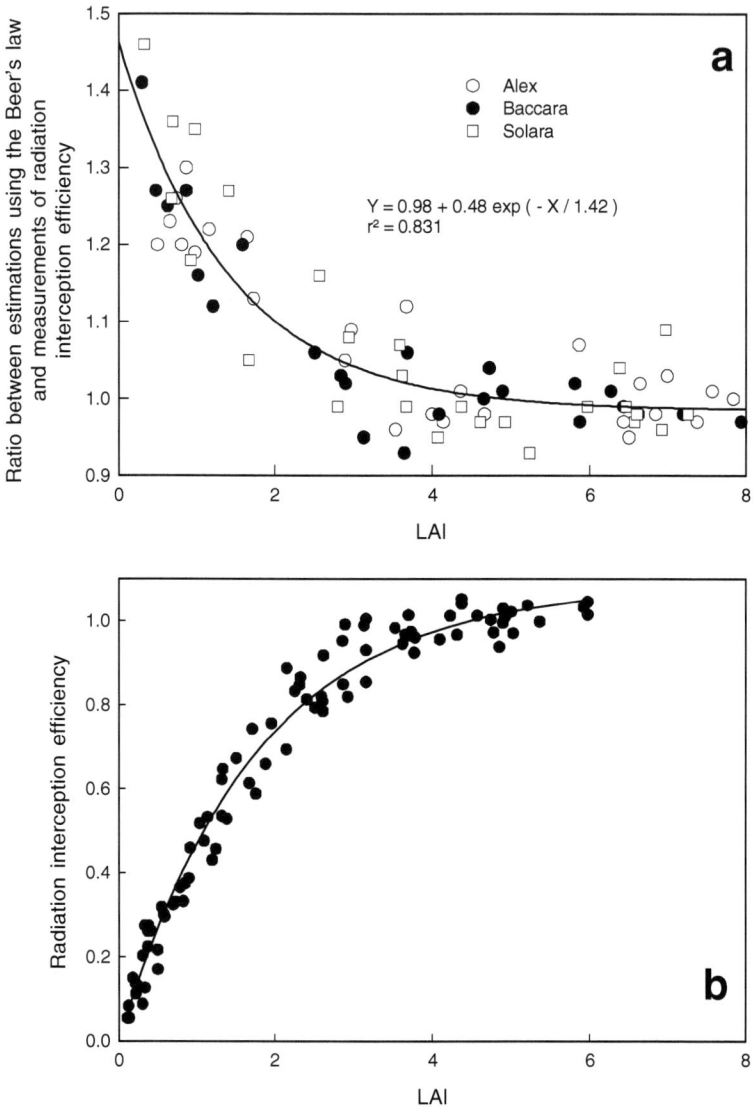

Figure 1.19. (a) Ratio between radiation interception efficiency estimated and measured as a function of the leaf area index of the canopy (LAI). Data are the result of a field experiment in Montpellier in 1996 (Lecoeur, n.p.). Radiation interception efficiency was estimated from LAI measurement and Beer's law (b) Radiation interception efficiency by the canopy (RIE) according to the leaf area index of the canopy (LAI). Data are the result of field experiments in Montpellier in 1996 and 1997 with several varieties, sowing dates and densities (see Lecoeur and Ney, 2003). The continuous line represents the fit using Beer's law and a light extinction coefficient k = 0.57.

plant phytomer by means of the allometric relationship between organ lengths and their area. By using a development model the number of phytomers with expanded leaves can be estimated in relation to thermal time. By knowing the number of stems per m², LAI variablility over the course of time can be obtained. In pea, LAI development over a cycle (and RIE by using equation 5) can be estimated by means of a single measurement of stipule lengths at the end of flowering (Lecoeur et al., 2001).

– Estimation of the radiation extinction coefficient

The radiation extinction coefficient (k) can be estimated by fitting Beer's law between LAI and RIE measurements. This estimation on a data set encompassing several varieties and densities gives a value of 0.57 (fig. 1.19b; see Lecoeur and Ney, 2003) which is close to the standard value proposed for leguminous plant canopies of 0.60 (Varlet-Grancher et al., 1989).

– Correspondence between RAE and RIE

Beers's Law can be applied by taking account of a maximal absorption efficiency.

$$RAE = RAE_{max} (1 - e^{-k\,LAI})\qquad\qquad\text{(Eq. 6)}$$

RAE does not exceed a maximal value which is constant for pea canopies with high densities. For these canopies, the maximal value is not determined by the LAI, but by the optical properties of the canopy. Ney (1994) reports a RAEmax value of 0.931 for the Solara variety. Similar values (0.924) have been measured in the Cheyenne variety (results from the winter pea French network of 2001–2002, np). In addition, comparative studies have shown that there were few differences between the extinction coefficients of leafy varieties and afila, in spite of very different leaf structures (Heath and Hebblewaite, 1985).

• By means of measurements of structural or optical characteristics of the canopy

Certain methods use "fish-eye" optical sensors which measure radiation in several directions of the area. The LAI-2000, LI-COR® is a case in point measuring radiation in blue (between 320 and 490 nm) in five directions of the hemisphere or when hemispherical photographs are used to recalculate probabilities of light-object contact and, in this way, to estimate a RIE.

By way of information, we can also report a RIE estimation method based on measuring canopy reflectance in two spectral bands (Ridao et al., 1996) whose principles and limitations are discussed by Sinclair and Muchow (1999).

Finally, computer-generated 3D plant representations can be used to estimate RIE or RAE by means of a digital radiative balance on these 3D objects. The 3D representations can be obtained by means of either plant digitalization, or a combination of organ development and expansion models. Using these approaches, for example, the impact on RIE of the differences in architecture between varieties can be tested (Rey *et al.*, 2001).

RUE

RUE is calculated (from equation 3 for example) and not measured. The quality of its estimation depends, therefore, on the reliability of the biomass estimations and efficiencies relating to radiation.

Variability sources of interception and absorption efficiency during the cycle

Change during the cycle

Interception efficiency changes during the cycle according to the leaf area index. When represented in terms of thermal time (fig. 3), the change in RIE is as follows: efficiency increases exponentially, then after a short transition, reaches a plateau. The value of the plateau depends on the maximum LAI reached by the canopy. A characteristic of pea is that RIE does not decrease at cycle end with leaf senescence. Pea leaves do not fall at senescence. They maintain their orientation and reduce their areas by only 5–10 % as they dry out. RIE thus remains close to plateau value until physiological maturity of the plant. The change in RAE is similar (Ney, 1994). For standard conditions of spring pea crops and for the main French production areas, Ney (1994) proposed a logistic curve to describe the relation between RAE and thermal time:

$$RAE = \frac{RAE_{max}}{1 + \exp\left(4Vm\left(P2 - \frac{CDD}{RAE_{max}}\right)\right)} \qquad (Eq. 7)$$

CDD represents cumulative degree-days from emergence, *P2* represents thermal time at the inflexion point, i.e. when RAE is approximately equal to 0.50. *Vm* corresponds to the rate of increase of RAE at inflexion point. The characteristic values of the parameters for spring pea are $RAE_{max} = 0.931$,

Vm = 0.00335 and $P2$ = 402. This relationship is only valid for dense canopies (densities equal to or greater than 80 plants/m²). The application of this same approach in winter pea runs up against the variability of the vegetative phase duration which does not allow for a single set of parameters to be proposed (Brun, 2002). The use of a dissymmetrical logistic could, in this case, better account for the winter vegetative phase which is quite long.

Effect of sowing date and density

The general rate of change of RIE and RAE during the cycle is the same regardless of the sowing date (fig. 1.20). A "latency" phase can always be noted where RIE (or RAE) almost does not increase, then a very rapid growth phase before reaching a plateau. Sowing date impacts only on the length of the "latency" phase. The equation 7, by considering that RIE max is equal to 1, makes it possible to account for the effect of the sowing date on the change in RIE. Vm is not modified by the sowing date. On the other hand, P2 increases with the earliness of sowing. It can reach 1400°Cd after sowing for October sowings.

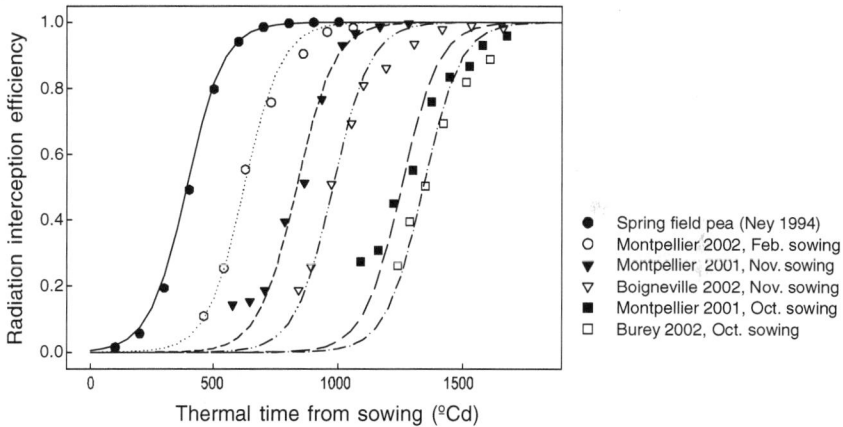

Figure 1.20. Changing patterns of radiation interception efficiency by the canopy (RIE) according to thermal time. The data come from various field experiments with densities of 80 plants per m² and a range of sowing dates from October to March (see Brun, 2002; Guilioni *et al.*, n.p.). The lines represent the fit using the formalism proposed by Ney (1994) which fixes RIE$_{max}$ at 1 and Vm at 0.00335 and modifies only P2. The values of the P2 parameters are 603, 822, 977, 1253 and 1340 for the curves starting on the left of the graph. The r² for all of the data is above 0.95.

Sowing density modifies the kinetics of crop LAI and the maximum LAI reached by the canopy. Thus, it has consequences on the date where RIE or RAE are at their maximum and on the value of this maximum (fig. 1.21). The plasticity of pea development, particularly in terms of the number of axes, enables it to reach the maximum value of RIE for densities equal to or greater than 40 plants per m².

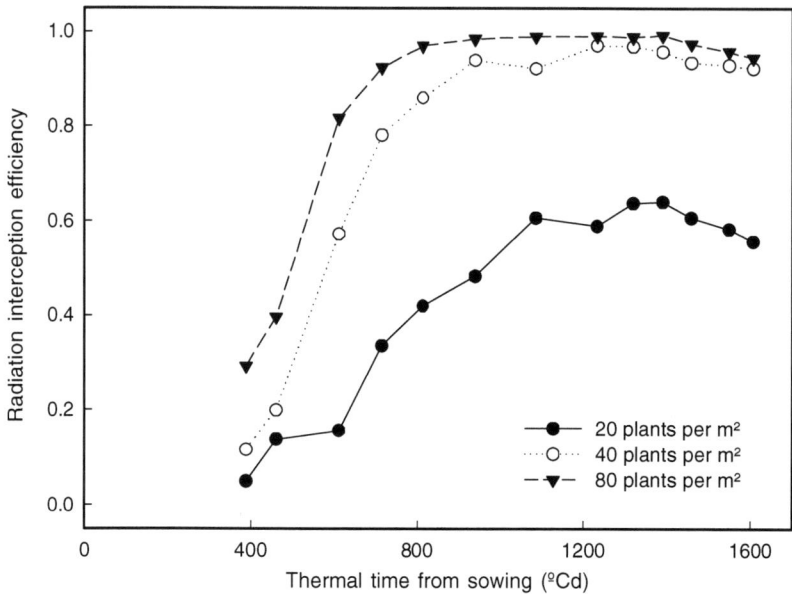

Figure 1.21. Changes in radiation interception efficiency by the canopy (RIE) according to thermal time since emergence: Solara variety in Montpellier in 1997, sown in February (see Lecoeur and Ney, 2003).

Genetic variability

At any given time, radiation interception efficiency by a canopy is the result of crop LAI, its spatial pattern and the optical properties of the vegetal organs which constitute it. During the cycle RIE depends on the development rate of the leaf area. Among the pea genotypes used in France, there is a great variability of these characters with, for example, leafy genotypes or afila, a greater or lesser predispostition for ramification, a more or less prominent indeterminate character, varied phyllochrons, etc. The combination of these characteristics has consequences on RIE, which are non-quantifiable by simple models. A potential method of quantification is the application of a radiative balance with 3D architectural representations of pea plants changing in relation to thermal time (Rey *et al.*, 2001).

Sources of variability of radiation use efficiency

Radiation use efficiency varies between species and especially between those which present different carbon metabolisms (C3 or C4). It also depends on the energy production cost of certain tissues or compounds. Thus, the production of an oleaginous seed is more expensive than that of a starch-rich seed. But, even within a species, there are many sources of variation. To our knowledge, little work has been carried out on RUE in pea. Values obtained in the literature in the absence of biotic, hydrous or nitrogen constraints vary between 1.25 g.MJ^{-1} of intercepted PAR (Pyke and Heddley, 1985) and 4.34 g.MJ^{-1} of intercepted PAR (Lecoeur and Ney, 2003). The periods of measurement, the studied genotypes, the sowing dates and crop densities generally account for the variability.

Variation during the cycle

Strong correlation coefficients (greater than 0.95) of the linear regression between produced biomass and intercepted PAR have often led to a constant average radiation use efficiency being considered over the cycle. However, the relationship is not strictly linear. Certain biomass values differ from the values estimated with a linear model (Lecoeur and Ney, 2003; fig. 1.22). This shows that RUE is not strictly constant during the cycle. By using experiments with water deficits, Martin et al. (1994) showed different changes in RUE before or after the beginning of flowering. Ridao et al. (1996) considered RUE to be a function of the physiological state of the plant and made a distinction between the phases before and after the beginning of seed filling. The beginning of seed filling corresponds to a more or less clear increase in RUE according to genotypes. In broad bean the increase in RUE during seed filling is explained by the reduction in nitrogen assimilation cost and by the increase in photosynthetic activity due to the presence of significant sinks (Husain et al., 1988, quoted by Ridao et al., 1996). Jeuffroy and Ney (1997) linked the changes in RUE to source-sink relationships within the plant and to pod and seed growth. They showed an increase in RUE during pod growth and a concomitant decrease with seed filling. Lecoeur and Ney (2003) refined the description of the change in RUE in relation to phenology and described the change in RUE according to thermal time since the emergence by a sinusoid (fig. 1.23). During the vegetative phase, RUE decreases from emergence until a minimal value reached before the beginning of flowering. Then RUE increases during the reproductive phase and reaches its maximum after the beginning of seed filling. It decreases then to zero corresponding with plant death. In the view of these authors, this change is a result of the sequential growth and senescence of the different plant organs (roots, stems, leaves, pods and seeds).

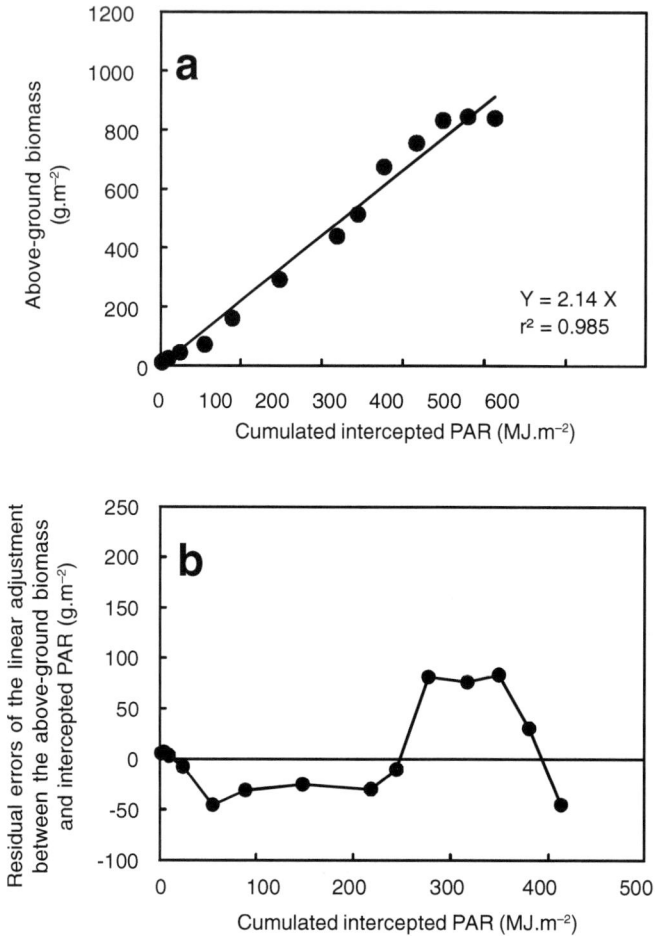

Figure 1.22. (a) Relationship between aerial biomass and PAR intercepted by a pea canopy, cv. Alex (b) relationship between difference on the linear model (measured dry matter-estimated by the linear model) and intercepted PAR (see Lecoeur and Ney, 2003).

Effect of sowing date and density

Lecoeur and Ney (2003) estimated RUE for spring pea canopies from 18 to 104 seedlings per m^2 with sowing dates spread over more than two months. Neither sowing density nor date had an effect on the sinusoidal rate of change of RUE during the cycle. The maximum readings and the thermal dates on which they were reached, on the other hand, were modified. Although the differences are not significant, the maximum readings of RUE

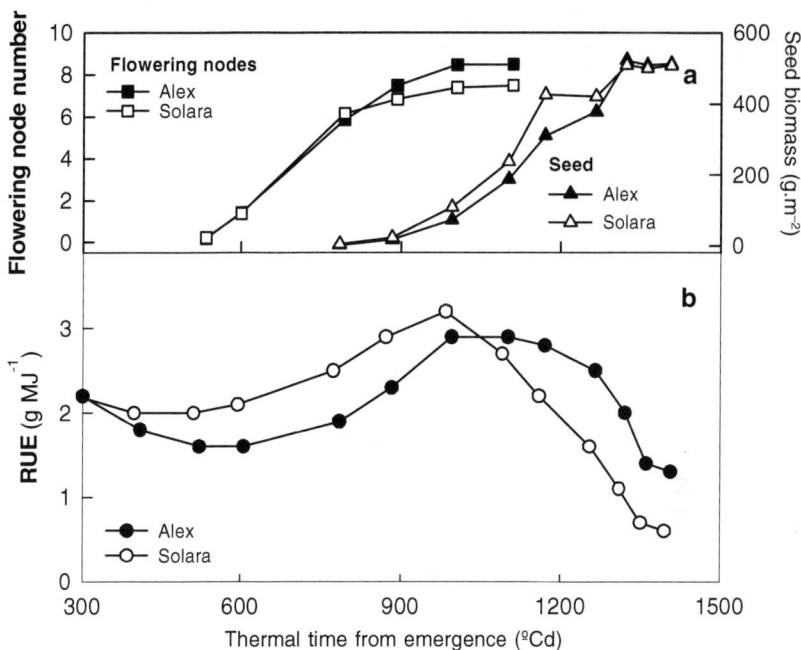

Figure 1.23. Example of genotypic variability of the radiation use efficiency (RUE) in pea. (a) Changes according to thermal time since emergence (a) of the number of flowering phytomers and seed biomass (b) of RUE. (see Lecoeur and Ney, 2003).

were all the greater, the thicker the density. These results correspond with those obtained by Ridao *et al.* (1996) which show an insignificant increase in RUE when sowing density rises from 80 to 100 seeds per m². In contrast, the latest sowings undoubtedly presented the weakest RUE maxima in relation to higher temperatures. Indeed, different sowing dates position the cycle over different periods of the year and thus subject the crop to varied climatic conditions, resulting in different levels of photosynthesis which can affect RUE (cf p. 154 and fig. 2.8).

Genetic variability

In pea, genotypic variability of RUE is of the same order as the variability of RUE between sowing dates or densities for the same genotype (Lecoeur and Ney, 2003). The comparison between genotypes is only meaningful if the experiments are conducted on the same site and at the same period to avoid confusion between the genetic effect and the effect of the environmental conditions.

When genotypes differ only by their leaf structure (leafless type, semi-leafless, leafy), there is no significant difference in RUE (Pyke and Hedley, 1985; Martin et al., 1994, Ridao et al., 1996). An experiment conducted in Dijon with seven genotypes showed a different rate of change of RUE according to genotypes, in connection with their development rate. The sinusoidal rate is maintained regardless of the genotype but an earlier seed filling leads to a RUE peak earlier in the cycle (fig. 1.23, Lecoeur and Ney, 2003).

Effects of environmental conditions

Water (Heath and Hebblethwaite, 1987) or nitrogen (Sinclair and Horie, 1989) deficits reduce RUE. These limiting factors and their effects on canopy physiology are addressed in the second part of this work. However, it is important to note that even a well-managed crop can experience sub-optimal conditions. Thus, uncontrollable environmental factors, such as radiation and temperature, can also affect the change in RUE during the cycle.

A host of theoretical studies (Choudhury, 2000), in which measurements were carried out on various species including leguminous plants, such as soybean (Sinclair et al., 1992) have shown that the increase in the proportion of diffuse radiation, allied with a drop in incidental radiation, increased RUE. Consequently, parcels of pea cultivated in regions differing by the proportion of diffuse radiation (overcast skies, fog) can present different RUE and, subsequently, different levels of biomass production.

RUE also varies with thermal conditions in connection with the response of photosynthesis to temperature. In pea, net photosynthesis is at its maximum for a leaf temperature between 15°C and 25°C. It is considerably reduced for temperatures below 15°C (Feierabend et al., 1992) and decreases in a quasi-linear way beyond (Guilioni et al., 2003; p. 154 and fig. 2.8)

Based on data in the literature and experiments with different sowing periods, a relationship has been proposed between RUE and average air temperature where RUE is expressed in relation to an observed maximum for average air temperatures of 18°C (Guilioni, 1997). RUE is reduced for average air temperatures below 12°C or above 22°C (fig. 1.24). This explains why late sowings, exposed to higher temperatures, present lower RUE values (Lecoeur and Ney, 2003) and a reduction in biomass production (Guilioni et al., 1998, p. 154 and fig 2.8). This could also explain why winter pea performances are lower than expected, even in the absence of frost. Autumn sowings are exposed to sub-optimal temperatures for RUE over almost the whole of the vegetative phase.

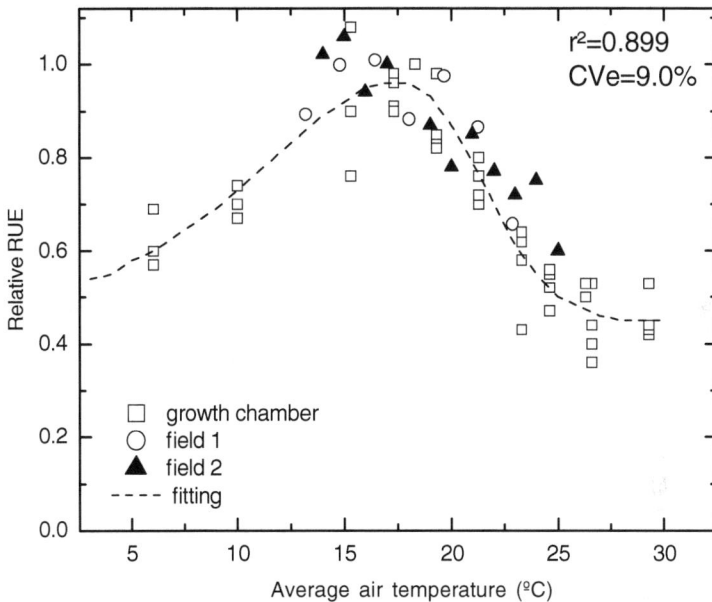

Figure 1.24. **Radiation use efficiency as a function of air temperature.** RUE is expressed relative to RUE measured at 18°C. Data from experiments conducted by several authors in controlled conditions (Stanfield *et al.*, 1966) and in the field (Guilioni, 1997).

Conclusion

In order to estimate the productivity of pea in fluctuating climatic conditions, a detailed knowledge is required of radiation interception (or absorption) efficiency and radiation use efficiency. There are simple and robust methods to estimate RIE. The use of portable devices such as LI-COR® LAI2000 gives rapid and reliable estimations of RIE as they operate on the basis of physical principles with few or no hypotheses on canopy structure. The robustness of Beer's approach also affords a reliable estimation of RIE based on the LAI measurement. The primary limitation is the cumbersome nature of this measurement.

The comparison of measured RIE with a reference represents a useful diagnostic tool of biomass production and yield limitation coupled with the installation of aerial sensors on a pea crop.

A global RUE value for the whole cycle is insufficient since it varies considerably over the season, as the various plant organs develop. RUE also responds to climatic variables like temperature and radiation. The main difficulty in estimating RUE thus derives from the interaction between the effects linked to phenology and environmental conditions. However, the definition of an average range of RUE over the cycle made it possible to improve considerably the evaluation of potential yields of this crop as well as accounting for abiotic constraints such as high and low temperatures and water deficit.

3

Dilution curve

Aurélie Vocanson, Nathalie Munier-Jolain,
Anne-Sophie Voisin, Bertrand Ney

For numerous species, plant N concentration decreases as biomass accumulates throughout the growth cycle, while N accumulation continues. This dilution of nitrogen by carbon assimilates accounts for the allometric relationship between N uptake and biomass accumulation (Salette and Lemaire, 1981). The decline of the N concentration in vegetative aerial organs over time may be attributed to a decrease in the leaf/stem ratio and to remobilisation of nitrogen from shaded to illuminated parts of the canopy (Lemaire *et al.*, 1992). The N concentration of the crop can be related to biomass production following a general equation presented below, which accounts for environmental and genotypic effects in non-limiting conditions:

$$\%N_{crop} = \alpha^* \, (DM_{crop})^{-\beta} \qquad \text{(Eq. 1)}$$

with $\%N_{crop}$ and DM_{crop} being the N concentration and biomass (in $t.ha^{-1}$) of the aerial parts of the crop respectively .

This relationship, which was named the "dilution curve", was fitted for grasses, especially cereals. On these species, this curve is valid during the vegetative growth cycle. Later it was used for lucerne and then grain legumes.

Legumes have specific traits that can modify the general pattern found in cereals. Indeed, growth of the pods carrying seeds with a high N concentration after flowering could compensate partly or wholly for the decline in N concentration in the vegetative parts of the plant.

Plant N concentration in grain legume crops in relation to developmental stage

The available data collected from legume crops show conflicting results among species:

- In pea, until biomass has reached approximately 1 t.ha^{-1}, N concentration increases with biomass, then decreases following the N dilution curve and finally remains stable from the beginning of the seed filling stage (fig. 1.25). These results show the determining role of the appearance of reproductive stages on the variation in N concentration of the crop during growth. Until emergence of the pods, plant N concentration declines as a consequence of the increase in biomass of vegetative organs as described previously for other crops. Then, after pod emergence and during seed filling, the accumulation of nitrogen in reproductive organs with a high N concentration counterbalances the decline in N concentration of the vegetative organs and therefore plant N concentration remains more or less constant.

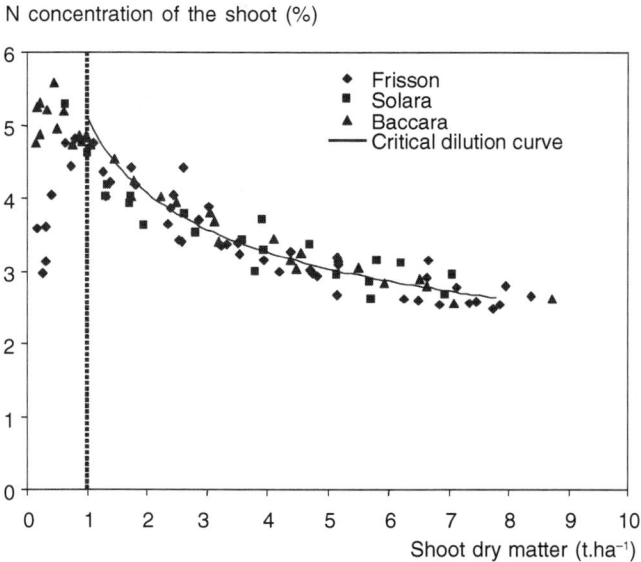

Figure 1.25. Dilution of nitrogen and critical nitrogen dilution curve until the beginning of seed filling stage in a pea crop. 3 genotypes: Frisson, Solara, Baccara.

- As shown by Lemaire and Allirand (1993) in lucerne, for which the "seed" sink is weak, plant N concentration decreases with biomass accumulation, as in cereals and in other forage crops.
- This is not the case in lupins (Duthion and Pigeaire, 1991) or soybeans (Ney *et al.*, 1997), as plant N concentration remains constant between flowering and the beginning of the seed filling stage, regardless of season, N nutrition regime, or cultivar for inoculated plants.

The behaviour of peas is intermediate between that of lucerne and soybeans. The influence of the developmental stage on changes in the plant N concentration of legume crops could account for the apparent diversity of behaviour observed among legume species. Lucerne, when cultivated as forage, does not produce seeds and therefore its plant N concentration follows the patterns described for cereals, whose seeds have a low protein concentration. In contrast, for soybean, that attains only 1t.ha^{-1} biomass by the beginning of flowering, plant N concentration remains constant until the beginning of the seed filling stage, due to the small proportion of vegetative biomass relative to that of the reproductive organs, which are rich in N. For peas, the situation is intermediate. Flowering occurs at a greater biomass than in soybean, which explains why N dilution in the biomass was observed at least until 5 to 6 t.ha^{-1}. From the beginning of seed filling, the decline in plant N concentration is halted by substantial N accumulation in pods and seeds, as in soybean. Thus, depending on the earliness of the flowering stage, plant N concentration will cease at different aerial biomass values and hence at different plant N concentrations according to the environment or plant species. The nitrogen dilution curve approach is therefore unreliable for species like soybean or lupin for which seed filling occurs early in the growth cycle.

The plant N concentration reached during seed filling depends upon the ratio between the reproductive compartment (which depends upon on seed number) and other compartments of the plant. Any stress which induces seed abortion will affect this ratio and therefore modify plant N concentration at the end of the growth cycle.

In pea, diagnosing nitrogen nutrition by the means of the nitrogen dilution curve seems to be valid for crop biomass above 1 t.ha^{-1} and until the beginning of seed filling. Thereafter, during seed filling, plant N concentration remains constant, as N dilution has ceased.

Determination of critical plant nitrogen concentration for pea

The critical N concentration curve corresponds to the level above which increase in plant N concentration will not lead to increase in biomass. Any point situated below this curve reflects a situation of sub-optimal N nutrition. Comparison with the critical N dilution curves thus provides a diagnosis of N deficiency (cf. p. 158).

For peas, symbiotic mutants were used to define the critical N dilution curve (Sagan *et al.*, 1993), following methods used for cereals (Justes *et al.*, 1994):

$$\%N_c = 5.08* (DM)^{-0.32}$$ Eq. 2

where $\%N_c$ and DM are the critical shoot N concentration and shoot biomass (in t.ha^{-1}) respectively .

Coefficients of the equation obtained for pea are very similar to those founds for lucerne (Salette and Lemaire, 1981; Lemaire *et al.*, 1985) and for C3 species in general (table 1.1).

Table 1.1. Values of α and β coefficients (cf. eq. 1) for different species.

	Critical %N	
	α	β
Wheat	5.29	0.44
Rapeseed	4.48	0.25
Lucerne	5.50	0.36
Pea	5.08	0.32
Average C3	5.10	0.34
Association	3.60	0.34
Maize	3.40	0.37
Average C4	3.50	0.35

Values of N concentration and biomass generally measured in the field are positioned on the dilution curve as defined above. This tends to show that in many situations, pea crops have the capacity to maintain their plant N concentration close to the optimum as defined by the dilution curve.

Determination of the « maximal » curve

In agricultural situations, it is not unusual to find pea crops with plant N concentrations above the critical level (Doré, 1992) as in numerous other plant species (Justes *et al.*, 1997 on wheat; Colnennet *et al.*, 1998 on rapeseed; Plénet and Guz, 1997 on maize; Duru *et al.*, 1997 on grassland). In certain situations, peas thus have the ability to absorb more nitrogen than the amount strictly needed to ensure maximal growth, as defined by the critical N dilution curve. Meynard *et al.* (1992) suggested that there is a maximal capacity of a pea crop to accumulate nitrogen in its shoot.

Following the method proposed by Justes *et al.* (1994) for winter wheat, a maximal dilution curve was established for peas using a data set comprising different spring and winter genotypes in situations where nitrogen was not limiting and where crop growth rate was maximal. As for the determination of the critical curve, coefficients of maximal curve were established during the period « total biomass > 1 t.ha^{-1} – end of flowering », that corresponds to the validity domain of the relationship:

If MS > 1 t.ha^{-1}, then %N$_{max}$ = 9.28 * (DM)$^{-0.55}$ (R^2 = 0.93) (fig. 1.26)

with %N$_{max}$ and DM being the maximal aerial N concentration and shoot biomass of the crop (in t.ha^{-1}) respectively.

The maximal curve is not shown after the beginning of seed filling in figure 1.26 as this period is outside the validity domain of the curve. Experimental points with N concentration above the critical N concentration correspond to situations where mineral N availability of the soil was very high, i.e. where symbiotic N fixation was low compared to mineral root absorption. While nitrate reduction takes place mainly in the roots of peas, when mineral N uptake by the roots is high, a large amount of nitrate is transported by the xylem sap to photosynthetic organs (Atkins *et al.*, 1980; Peoples *et al.*, 1987): the reduction of nitrate then occurs in the leaves where it is coupled to photosynthetic reactions. This results in lower carbon costs than when it occurs in the roots. This mechanism, which saves energy, would explain the increased plant N concentration for a given biomass.

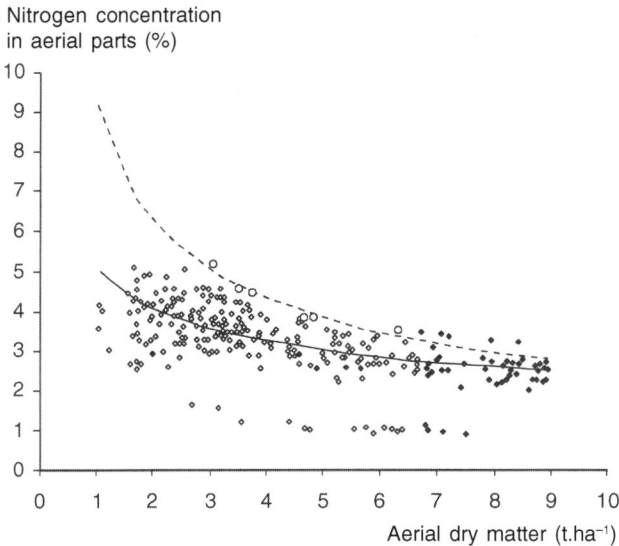

Figure 1.26. N dilution curve in biomass in a pea crop: critical curve—(full line) (% No) and maximal curve—(dotted line) (% N$_{max}$), white diamonds are for crops in the vegetative period (before flowering), black diamonds are for crop at the beginning of the reproductive period (between flowering and the beginning of seed filling). White circles are experimental points used to build the maximal curve.

However, this first estimate of the maximal curve for a pea crop during the vegetative phase requires verification on a larger data set to improve its robustness.

Hypothesis for the control of nitrogen uptake by the plant

The stability of the relationship defining the critical dilution curve suggests that N uptake is controlled by plant growth, that defines N demand (Lemaire *et al.*, 1997). This hypothesis has been reinforced by recent studies that show that internal factors produced by the shoot (especially molecules originating N metabolism) act as retro inhibitors on N uptake by the roots (Muller and Touraine, 1992) and by nodules (Soussana and Hartwig, 1996; Neo and Layzell, 1997). However, high soil nitrate availability can lead to mineral N uptake by the root in larger amounts than those defined by N demand. These observations suggest that root absorption is also controlled by soil nitrate availability.

Thus, considering observations made on non-leguminous plant, the control of nitrate uptake in the field would depend on the nitrogen status of the plant, as determined by the nitrogen nutrition index (NNI) (Devienne *et al.*, 2000). Below the critical nitrogen dilution curve (NNI<1), mineral N uptake would be essentially controlled by shoot growth potential and by nitrate availability, that determines the actual growth of the plant. Above the critical nitrogen dilution curve, when growth is maximal, N uptake would be controlled by soil mineral N availability only. The plant then stores nitrogen in the form of nitrate and proteins, and its nitrogen nutrition index is greater than 1.

There could be a similar mechanism in legumes, bearing in mind that when soil nitrate availability in low, N uptake mainly originates from symbiotic fixation of atmospheric nitrogen.

Root and nodule establishment and associated carbon costs

Anne-Sophie Voisin, Christophe Salon, Yves Crozat

Yield and seed protein content of legumes strongly depend upon nitrogen nutrition. In these species, N acquisition by the plant relies both on root nitrate assimilation and N_2 fixation through symbiosis with *Rhizobium* bacteria. While symbiotic bacteria have the ability to fix atmospheric N_2, the host plant provides them with the carbohydrates they need. Symbiotic fixation and root assimilation are complementary for optimal nitrogen nutrition (cf. Chapter: "Nitrogen nutrition efficiency") but may be antagonistic for carbon use within the root system: (i) for roots and nodules establishment (structural components and respiration for associated synthesis processes), (ii) for organ maintenance (maintenance respiration) (iii) and for their function. A better understanding of roots and nodules establishment together with associated carbon costs due to organ establishment, maintenance and function is therefore a prerequisite for improving complementarity between the two nitrogen acquisition pathways of legumes without inducing additional costs at the whole plant level.

Root and nodule establishment

Establishment of the roots

The root system of legumes is made up of a taproot and first and second order lateral roots which appear acropetally on the taproot. Root development and growth can be divided into three phases (Mitchell and Russel, 1971):

- **Taproot elongation:** during early rapid vegetative growth of the shoot, the rapid downward taproot development is accompanied by horizontal lateral root development primarily in the upper 10 cm of the profile. In peas, the elongation rate of the taproot is maximal between germination and seedling emergence; then it declines roughly in conjunction with the exhaustion of seed reserves (Tricot et al., 1997). It seems that the taproot has a higher priority for carbon assimilates than lateral roots as the decrease in assimilate availability—due to the transition from heterotrophy to autotrophy for C—first affects lateral roots (by reducing the density and length of first order laterals) and then the taproot by reducing its elongation rate (Tricot et al., 1997).
- **Rapid increase of root biomass:** during flowering and pod formation, high rates of root growth are associated with rapid development of second order lateral roots and lead to deeper penetration of the root system into the soil. In peas, maximum rooting depth is generally observed at flowering and varies from 50 to 80 cm as a function of genotype and soil and weather conditions. (Hamblin and Hamblin, 1985; Thorup-Kristensen, 1997).
- **Slowing down of root biomass accumulation and emergence of the last lateral roots:** this takes place during the seed filling phase. Several authors have suggested that root growth stops at flowering (Salter and Drew, 1965). However, others have shown that root biomass of peas increases later in the growth cycle, until

physiological maturity of the seeds. (Mitchell and Russel, 1971; Armstrong and Pate, 1994).

The root system of legumes is less extensive than that of cereals (Greenwood *et al.*, 1982; Hamblin and Tennant, 1987). At every developmental stage, most of the roots (up to 90 %) are located in the upper soil layer (Hamblin and Hamblin, 1985; Armstrong *et al.*, 1994).

At physiological maturity, for the same biomass, the root length of legumes is only half that of cereals. It seems that legumes have adopted a more opportunistic strategy as they first establish a rather modest root system; however it has the ability to increase later according to the water availability in the soil. Conversely, cereals always develop a fully branched root system, whatever the environmental conditions.

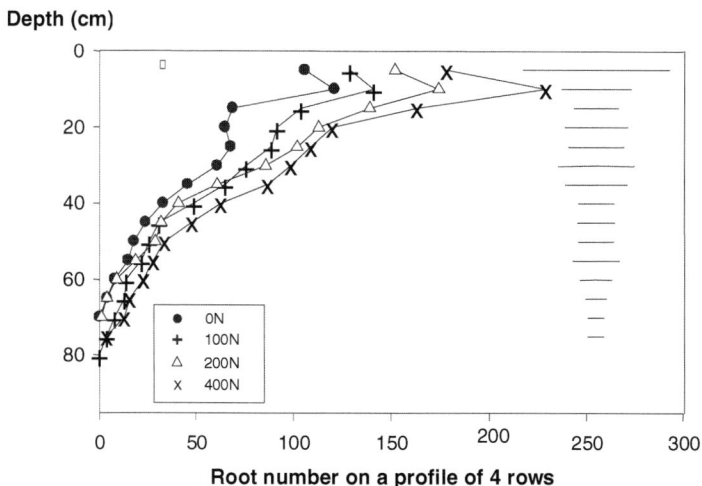

Figure 1.27. Root profiles obtained by root impact counting on 4 rows of peas at physiological maturity, as measured in a field experiment with mineral N supply at sowing varying between 0 and 400 kg N.ha^{-1}. After Voisin *et al.*, 2002.

In peas, soil structure is one of the environmental factors that most strongly affect the establishment of the root system. The dynamics of colonisation by and distribution of the roots in the soil profile are strongly modified when the soil is compacted. (cf. "Effects of compacted soil structure" p. 162). Locally, nitrate stimulates root proliferation. (fig. 1.27), which results in an increase in the "exchange surface" of the roots, mainly in the ploughed layer (the upper 20 cm). However, nitrates have no effect either on root distribution within the profile or on maximal rooting depth. Thus, the root system of peas remains shallow whatever the nitrogen nutrition regime.

• Nodules establishment and its modulation

At the organ scale

Pea nodules arise from cells adjacent to the root pericycle, following a complex sequence of mechanisms involving signal exchange and recognition of the symbiotic partners. The bacteria (*Rhizobium leguminosarum bv. Vicae*) that live saprophytically in soil are attracted towards the plant rhizosphere by exudation of various molecules, such as flavonoids Aggregation of the *Rhizobia* at the tip of the root hairs induces their deformation as a "shepherd's crook" and the bacteria invade the plant by a newly formed infection thread growing through it. Simultaneously, root cortical cells start differentiating, giving rise to a nodule primordium. Pea nodules are indeterminate (Brewin, 1991). The apical meristematic zone is persistent and is followed towards the central cylinder of the root by successive cellular layers that are representative of successive developmental stages of the nodule: proliferation of the infection thread, colonisation of cells by bacteroids (differentiated and active forms of bacteria) and senescence. As the senescence zone progressively dominates the other cell layers, the nodule falls away.

At the root scale

The infections leading to nodule formation are restricted to a narrow band of cells above the zone of root elongation and just below the smallest emergent root hairs (Bhuvaneswari *et al.*, 1980). This zone is only temporarily sensitive to infection (Pueppke, 1983). The duration (expressed in days) between infection by Rhizobia and nodule appearance varies with root temperature (Tricot, 1993). This duration is constant when expressed in cumulative temperature (threshold 6°C). Therefore, the date of appearance of the first nodules does not depend on the stage of foliar development (table 1.2).

Table 1.2. Foliar developmental stage and duration of formation of the first nodules formed on the taproot and on lateral roots as a function of root temperature. (After Tricot, 1993)

Root temperature (°C)	Leaf number when the first nodule appeared		infection-appearance duration (cumulative T°, threshold 6°C)	
	Taproot	Lateral roots	Taproot	Lateral roots
18	4.2 +/− 0.5	6.6 +/− 0.9	101 +/− 14	105 +/− 21
12	6.1 +/− 0.7	7.4 +/− 0.8	106 +/− 20	103 +/− 24

At the root system scale

Nodule appearance is characterized by "nodulation waves" spaced out in time; nodule distribution within the root system is uneven (Sagan and Gresshoff, 1996). Nodule density is maximal at the taproot base and then diminishes towards the apical zone to a zone bare of nodules (Tricot, 1993). In the same way, the proportion of lateral root carrying nodules is maximal (90 to 70 %) at the taproot base and strongly decreases as the distance from the taproot base increases (Tricot, 1993). Consequently nodules are mainly located in the upper part of the root system.

Tricot et al. (1997) observed that the probability of a nodule appearing on a given root segment increases with the segment's elongation rate. Root segments or laterals that are deprived of nodules are those whose elongation rate is above a threshold level (Tricot, 1993). These results suggest that the appearance of nodules and their distribution depend upon assimilate availability. Assimilate availability (C or N) could lead to a systemic control of nodule number (Atkins et al., 1989).

At the whole plant scale

Intrinsic regulation mechanisms (Cantano-Anollès and Gresshof, 1990, 1991), termed auto-regulation, have been suggested to account for the reduction of nodulation success following the appearance of new nodules (Kosslak and Bolhool, 1984). The number of newly formed nodules is also influenced by environmental factors (Mengel, 1994). In particular, nitrates decrease nodulation (Streeter, 1988). They act at all stages involved in the formation of functional nodules by alteration of flavonoid production (Bandyopadhyay et al., 1996), impairing root hair infection (Dazzo and Brill, 1978), or by blocking nodule development (Bauer, 1981). Regulation of nodulation by nitrate could involve a systemic control of the shoot on nodules, probably in synergy with auto-regulation mechanism (Francisco and Akao, 1993). Thus, the higher the nitrate availability, the lower is the nodule biomass.

Dynamics of nodules establishment over time

Once the first nodules have appeared, nodule biomass increases rapidly to reach a maximum at flowering, after which it remains constant until the final stage of seed abortion before steeply decreasing at the end of the growth cycle (Pate, 1958; Tricot-Pellerin et al., 1994). When senescent nodules are discarded, the biomass of active nodules is a good indicator of the fixing capacity of the plant (fig. 1.28). In this example, fixation increases until the 9th leaf stage and then slows down with active nodule biomass.

The overlapping of phases of determination of (i) nodule number and (ii) of their individual growth, makes it difficult to study the effects of a trophic constraint on the different components of the total biomass of the fixation apparatus (Crozat, 2000). In numerous situations, variability in the number of nodules is the major cause of variation of total nodule biomass (Crozat, 2000). However, when a trophic constraint is applied after the phase of elaboration of nodule number (after the beginning of the autotrophy phase in the case of the taproot, after flowering for the entire root system), this constraint mainly affects individual nodule growth or their degree of senescence (Tricot-Pellerin *et al.*, 1994; Lawrie and Wheeler, 1974).

Relative values (%)

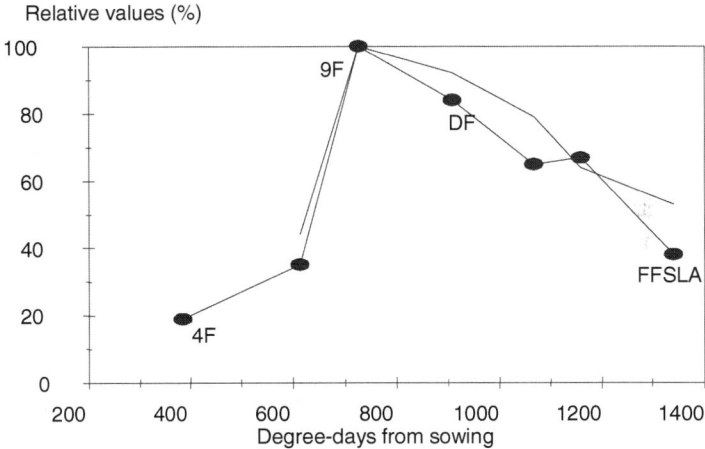

Figure 1.28. Comparison of time courses of active nodule biomass per plant (•) N fixing capacity, (indicated by acetylene reduction activity (ARA)) (○) as a function of thermal time (threshold 0°C). Values are relative (i.e. as % of the maximal value observed at the 9th leaf stage). After Crozat , 2000.

Carbon costs associated with structure establishment, maintenance and metabolism

Symbiotic N$_2$ fixation and nitrate root absorption require carbon for the morphogenesis and maintenance processes of associated organs (roots and nodules) as well as for the reducing activities of atmospheric or mineral nitrogen and for the transport of nitrogen compounds to the shoot.

- Carbon fluxes associated with the establishment and operation of the nodulated root system : C partitioning between shoot and root parts

Carbon uptake by the plant: At least until flowering, net photosynthesis does not depend upon the nitrogen nutrition regime (i.e. symbiotic fixation vs. mineral uptake) (Voisin *et al.*, 2002). However, during the seed filling period, net photosynthesis may be higher when the plant assimilates soil mineral N.

Variations in carbon partitioning between the shoot and the nodulated roots: The amount of photoassimilates transferred to the nodulated roots varies according to the legume species, the age of the nodulated roots and plant phenology (Pate and Herridge, 1978). In peas, it decreases from 45 to 7 % of net photosynthesis (Voisin *et al.*, 2003a) between early vegetative stages and the seed-filling period. However, for a given nitrogen nutrition level, the amount of carbon allocated to the nodulated roots is unaffected by the nitrogen nutrition source (symbiotic fixation or root absorption). This being so, nitrate availability does not affect the phloem stream of carbon towards to the nodulated roots (Voisin *et al.*, 2003a).

Partitioning rules of photoassimilates between the shoots and the roots were determined by analysis of variations in carbon partitioning between the shoots and the roots with varying photosynthesis rates.

- At the vegetative stage, the nodulated roots take priority over the shoots for photoassimilates .
- From flowering, the root's priority over the shoot decreases, owing to the appearance of reproductive organs which are a very competitive sink for assimilates.
- During the seed filling period, despite the low sink strength of the root parts (7 % of net photosynthesis), the level of priority for carbon of the nodulated roots equals that of the shoots.

Moreover, the amount of carbon allocated to the nodulated roots increases when the level of net photosynthesis increases. This indicates that the nodulated root sink remains limited by carbon at all stages of growth (Voisin *et al.*, 2003a).

- ## Carbon use within the root system as affected by the form of nitrogen nutrition

Root and nodule respiration
For a given nitrogen nutrition level, photosynthesis and carbon allocation to the nodulated roots are unaffected by the nitrogen nutrition regime (symbiotic fixation vs. root absorption). However, plants relying on both N_2 and mineral N for their nitrogen nutrition generally have a higher root biomass compared to plants relying only on symbiotic N_2. This can be explained by differences in carbon use efficiency within the root system. Respiration is the main component of carbon used within the root system. The fraction of C allocated to the nodulated roots that is lost by respiration is constant throughout growth but varies with the nitrogen nutrition regime as it is on average slightly higher for plants relying only on symbiotic fixation (83 %) compared with plants which are also absorbing mineral N (71 %) (Voisin *et al.*, 2003a).

Growth and maintenance of roots and nodules

The carbon respired by the nodulated roots is that used for synthesis of root and nodule structures, for their maintenance (due to the turn-over of compounds) and for nitrogen uptake activities (symbiotic fixation, root absorption). Of these, synthesis of structure has the highest respiratory cost. The cost of carbon synthesis probably limits both associated nodulated root growth and nodule N_2 fixing activity. However, the fraction of the synthesis cost of the total carbon use by nodulated roots decreases during growth. In spite of the higher maintenance cost (or turn-over) of nodules compared to roots, the total use of carbon at the whole plant level for synthesis and maintenance of the nodulated roots does not depend on the ratio of root and nodule biomasses (Voisin *et al.*, 2003c). Therefore, within the nodulated roots, nodule establishment is at the expense of root growth. Thus, nitrate availability results in higher root biomass because it limits nodule establishment (fig. 1.29).

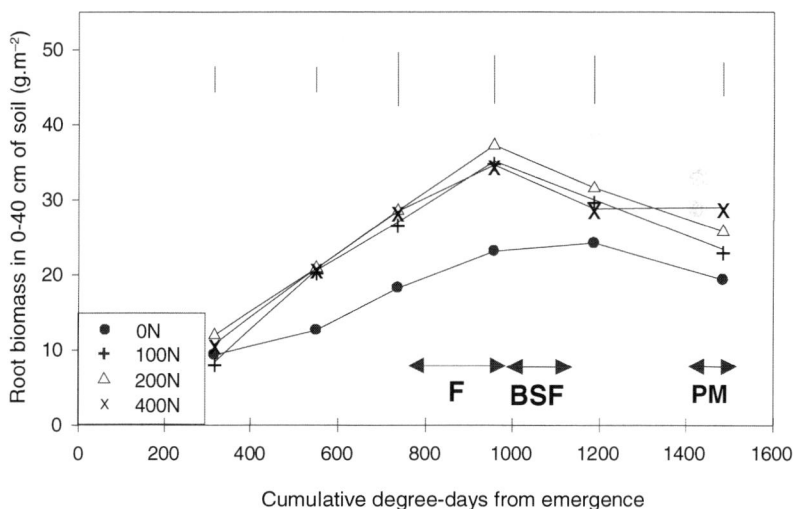

Figure 1.29. Time course of root biomass growth in the 0–40 cm soil layer during growth, from measurements in a field experiment with mineral N supply at sowing varying between 0 and 400 kg N.ha[-1]. (F: Flowering; BSF: Beginning of Seed Filling; PM: Physiological Maturity). After Voisin *et al.*, 2002.

At the whole plant level, nodule growth probably takes places at the expense of both shoots and roots during the vegetative stage, but only of roots at flowering (Voisin *et al.*, 2003b). These results explain why the shoot biomass is slightly higher for plants grown in the presence of nitrate compared to those relying only on symbiotic N fixation for their nitrogen nutrition. (Butler and Ladd, 1985).

During the seed filling period, root and nodule growth is probably complete. The carbon use of the nodulated roots is largely devoted to the maintenance of their structure. Therefore, the low sink strength of the root parts at the end of the growth cycle leads to more or less rapid senescence of the nodulated roots. At this late stage, it seems that carbon is preferentially allocated towards organs involved in the major nitrogen uptake pathway, that is roots in the presence of nitrates and nodules for plants relying on N_2 fixation (Voisin *et al.*, 2003b).

Prospects

Although a well-developed root system (induced here by the presence of nitrate in the soil) has no effect on yield in non-limiting weather conditions, a better root exploration of the soil could be an advantage for the plant, especially when water availability is low or when conditions are unfavourable to symbiotic fixation (compacted soil, excess salinity etc.) However, the effects of such a root system on the water and mineral uptake capacity of the plant remain to be tested. Therefore, creation of genotypes with nodule and root establishment rates adapted to environmental constraints constitutes an interesting challenge for breeders.

Nitrogen nutrition efficiency

Anne-Sophie Voisin, Christophe Salon

Nitrogen nutrition of legumes relies on both symbiotic fixation of atmospheric N_2 and soil mineral N absorption by the roots. The efficiency of nitrogen nutrition thus involves the regulation of different processes of root and nodule development (cf. Chapter: "Root and nodules establishment and associated carbon costs") together with regulation of the operation of the two nitrogen uptake pathways. Optimising the interaction between N_2 fixation and mineral N absorption throughout the growth cycle will result in (i) optimal nitrogen nutrition throughout growth, (ii) better use of soil mineral N through its transfer to harvested organs, (iii) a decrease in N residues in the straw at harvest.

Contribution of the two nitrogen uptake pathways to total N uptake by the plant during growth

- Complementarity between symbiotic N_2 fixation and mineral N root absorption

In legumes, the higher the nitrate availability, the lower is the contribution of symbiotic N_2 fixation to the total N uptake by the crop, symbiotic fixation being replaced by mineral N root absorption (fig. 1.30).

In pea, in non-limiting weather conditions, nitrogen requirements can be met whatever the nitrogen nutrition regime, as defined by the relative contribution of symbiotic fixation and root mineral N absorption (Sagan *et al.*, 1993, Voisin *et al.*, 2002a): in the absence of nitrate, nitrogen fixation can ensure an adequate nitrogen supply to the plant. (fig. 1.31). However, in soybean, nitrogen nutrition can be limiting if no fertiliser N is applied at sowing (Crozat *et al.*, 1994).

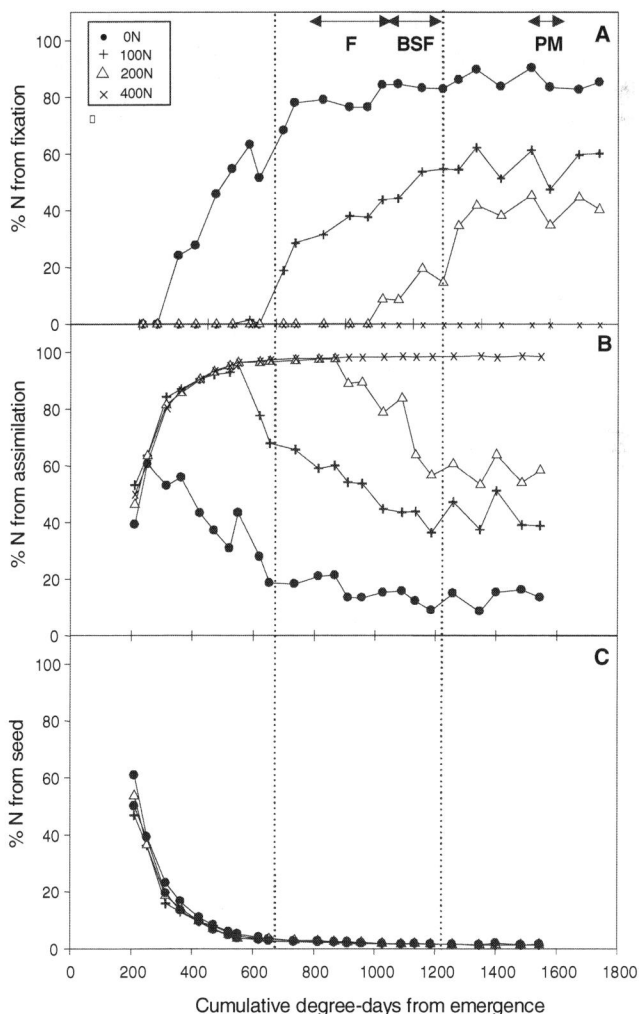

Figure 1.30. Time course throughout growth of N from symbiotic N_2 fixation (A), mineral root absorption (B) and from the seed (C), as measured in a field experiment with mineral N supply at sowing varying between 0 and 400 kg N.ha^{-1}. After Voisin *et al.* (2002).

Whatever the nitrogen nutrition regime, nitrogen uptake seems to be governed by nitrogen demand, as determined by plant growth rate, according to the critical nitrogen dilution curve (Ney *et al.*, 1997). It seems that nitrogen uptake is modulated by feedback control of the roots by the shoot through carbon and nitrogen fluxes (or "signal" molecules) circulating in the phloem. Regulation does not depend on the nitrogen source (root absorption or symbiotic fixation, Lemaire *et al.*, 1997) according to hypotheses for the regulation of both symbiotic fixation (Soussana and Hartwig, 1996; Neo and Layzell, 1997) and mineral N root absorption (Muller and Touraine, 1992). However, when the plant only relies on mineral N for its nitrogen nutrition (fig. 1.31, 400N treatment and 200N treatment at the beginning of the growth cycle) there are situations of luxury consumption of nitrogen, as shown by their position on the maximal dilution curve (cf. chapter: "Dilution curve" p. 61).

Nitrogen concentration in the shoot (%)

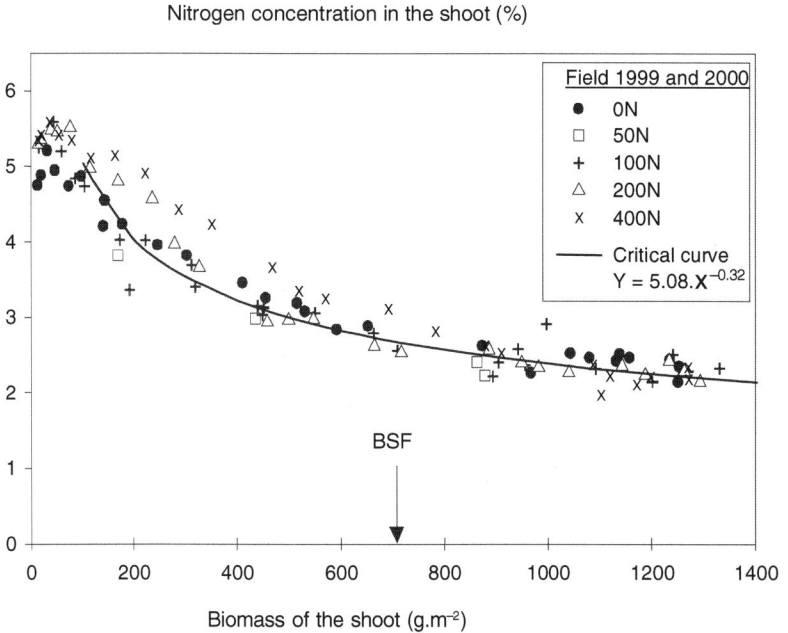

Biomass of the shoot (g.m⁻²)

Figure 1.31. Nitrogen content of the shoot as a function of its biomass, as measured in field experiments with mineral N supply at sowing varying between 0 and 400 kg N.ha⁻¹. After Voisin *et al.* (2002a).

- The two nitrogen uptake pathways have different kinetics and different sensitivities to environmental constraints

The two nitrogen uptake pathways contribute differently to total N uptake in space and time (fig. 1.30).

In the presence of nitrate in the soil

In the field, as long as soil nitrate availability remains above about 60 kg N.ha^{-1}, symbiotic fixation is inhibited and mineral N root absorption is the only possible pathway for nitrogen uptake by the plant. In agricultural conditions, when the amount of nitrate left in the soil is high, mineral N root absorption can greatly contribute to total N uptake by the plant. Later on, nitrate root absorption decreases as the soil mineral N pool becomes exhausted. Meanwhile, the inhibition of symbiotic fixation is lifted.

At the beginning of the growth cycle

At the beginning of growth, a slight and temporary nitrogen deficiency can occur when plants are grown without mineral N and thus rely on symbiotic fixation for their nitrogen nutrition (fig. 1.31). This can be explained by a lag phase in the establishment of fixing activity and by the transition from heterotrophy to autotrophy. This period of nitrogen deficiency is the result of carbon competition processes between nodules, roots and shoots for their establishment and metabolism. Differences between species in the duration of the initial lag phase can be accounted for by (i) the size of the seed, that determines its reserve pool and/or (ii) the germination type (hypogeal in pea; epigeal in soybean) which influences the distance between sources (the seed) and sinks (shoots and roots) and thus determines preferential allocation of the seed assimilates during the heterotrophic phase (Sprent and Thomas, 1984). This is why a small fertiliser N supply at sowing can have a positive "starter" effect in soybean (Starling et al., 1998). Nitrogen deficiency is however temporary. In pea, the presence of nitrate slightly stimulates vegetative growth at the beginning of the growth cycle but has no effect on seed yield and seed N content at harvest (Jensen, 1986).

During the growth cycle

Once the nodules begin to grow, symbiotic fixation allows the plant to substantially accumulate nitrogen, but nitrate may be absorbed by the roots if it is available (Vessey, 1992; Bergersen et al., 1992). At least until the beginning of seed filling, plants relying on various nitrogen nutrition regimes have similar levels of photosynthesis, growth rate and N uptake rates.

At the end of the growth cycle

Pea has the ability to accumulate nitrogen late in the growth cycle, although the amounts taken up may vary according to environmental conditions (Jensen, 1986; Beck *et al.*, 1991; Salon *et al.*, 2001). For most legumes, symbiotic fixation sharply decreases after flowering, either due to competition for assimilates with growing seeds or to environmental factors (Hocking and Pate, 1978; Lawrie and Wheeler, 1974; Sparrow *et al.*, 1995; Vessey, 1992). Indeed, in conditions of water stress (Hardin and Sheehy, 1980; Serraj *et al.*, 1999) high salinity (Sprent *et al.*, 1988) or phosphorus limitation (Drevon and Hartwig, 1997), symbiotic nitrogen fixation is reduced and mineral N root absorption assumes a major role in exogenous accumulation of nitrogen, even though it may be limited by the shallow root system of the pea crop. Despite being less sensitive to environmental conditions, mineral N root absorption also decreases during seed filling.

At the end of the growth cycle, it seems that symbiotic fixation is less efficient than nitrate root absorption. Indeed, higher levels of photosynthesis were measured in the presence of nitrates. Thus nitrate uptake at the end of the growth cycle may delay remobilisation of nitrogen from vegetative parts to the seeds and thus delay the self-destruction of the photosynthetic apparatus (Sinclair and de Witt, 1976). It may thus extend the duration of seed filling (cf. Chapters: "Individual seed weight" p. 109 and "Seed protein concentration" p. 117). Moreover, the amounts of nitrogen left in the soil after harvest are often high (Munier-Jolain and Carrouée, 2003), due to the modest root system of pea compared to cereals (Hamblin and Tenant, 1987).

> To summarize, the presence of nitrate in the soil delays nodule establishment and thus impairs the nitrogen fixing activity of the plant. Once the nodules are established, when mineral N is available, it seems that root absorption can complement or replace symbiotic fixation either when demand for N is high (at the end of the growth cycle for example) or when adverse environmental factors prevent symbiotic fixation from meeting the plant's demand for N.

N_2 fixing activity as modulated by growth, phenology and nitrate

• N_2 fixing activity as modulated by nitrate

Nitrate is the main environmental factor that limits fixing activity throughout the growth cycle (Streeter, 1988):

- • A general relationship was established between the percentage of nitrogen fixed throughout the growth cycle and the nitrate content of the ploughed layer at sowing (fig. 1.32). The decreasing linear relationship indicates that symbiotic fixation is inhibited in

proportional to the amount of nitrate available in the soil. Nitrate inhibition of symbiotic nitrogen fixation is absolute when soil nitrate availability at sowing is over 380 kg.ha[-1].
- Another relationship was established between the "instantaneous" percentage of symbiotic fixation measured at weekly intervals and

N from fixation (%)

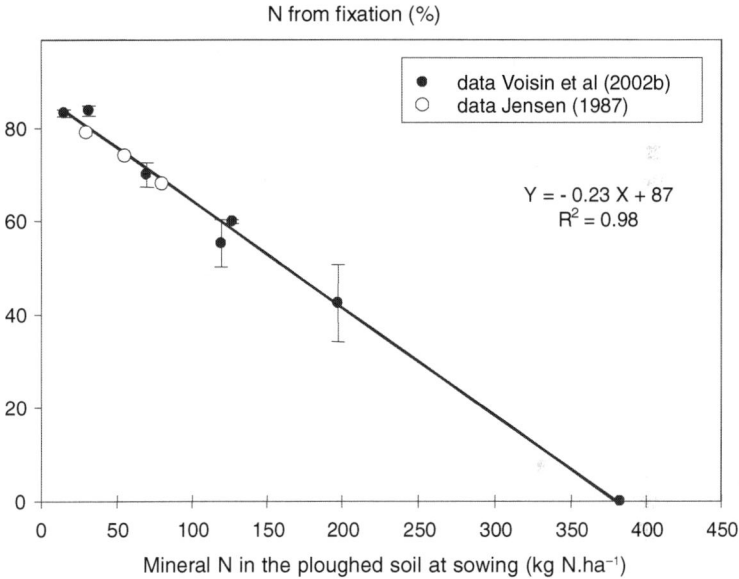

Figure 1.32. Contribution of symbiotic fixation to total N uptake by the plant throughout the whole growth cycle, as a function of nitrate availability in the ploughed layer at sowing. After Voisin *et al.* (2002b).

soil nitrate content during the same period (fig. 1.33). The percentage of instantaneous symbiotic fixation decreases linearly with soil nitrate content, the limitation being complete beyond a given threshold. Two linear relationships were established, depending on the plant growth stage. The first relationship is valid from sowing to the beginning of flowering, while the second applies from flowering to the beginning of seed filling. For a given level of soil nitrate content, soil limitation by nitrates is higher at the reproductive stages compared to the vegetative stages.
- These relationships were established in the field and are apparent relationships between symbiotic fixation and nitrate availability. The « mode of action » of nitrate and changes in fixing activity during the growth cycle must be accounted for by considering both nodule establishment and their fixing activity.

% of instantaneous fixation

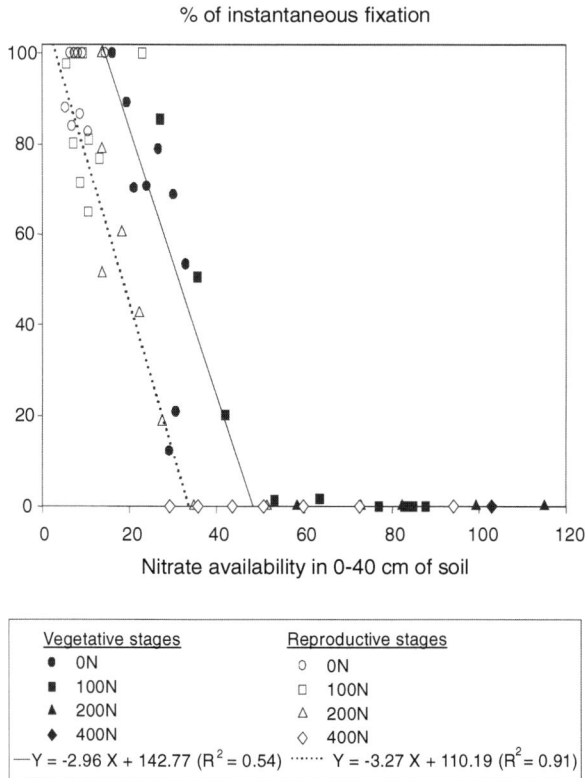

Figure 1.33. Percentage of instantaneous fixation as a function of nitrate availability in the soil as measured in a field experiment with mineral N supply at sowing varying between 0 and 400 kg N.ha^{-1}. Black symbols represent plants harvested during the vegetative period while white symbols are for plants harvested between flowering and the beginning of seed filling. After Voisin *et al.* (2002b).

- ## Time course of N$_2$ fixing activity during the growth cycle (fig. 1.30)

In field experiments (Voisin *et al.*, 2002b), when no fertiliser N is applied (with residual mineral N in the ploughed layer at sowing of around 30 kg N.ha^{-1}) symbiotic N$_2$ fixing activity starts after a 235 °C day delay from seedling emergence, which can be accounted for by a delay in the establishment of the nodules and onset of their activity. The initial lag phase increases with nitrate availability in the field because symbiotic fixation is totally inhibited as long as this is above 56 kg N.ha^{-1}. During growth, when soil nitrate availability falls below this threshold (due to mineral N uptake by the roots), the inhibition of nodule establishment is lifted. A longer lag phase can be observed when the threshold is reached after the beginning of flowering, presumably due to increased competition for carbon between nodule establishment and the formation of reproductive organs. It seems that nodule establishment can no longer be initiated after the beginning of seed filling.

Once symbiotic fixation has been initiated, symbiotic fixation rate per unit thermal time (mg N fixed per degree-day) increases linearly with thermal time, the symbiotic fixation rate accelerating as the nodules become established. The kinetics probably depend upon carbon availability for nodule establishment and therefore on both crop growth rate and carbon competition with reproductive organs. During this phase of nodule establishment, the level of fixing activity is related to nodule biomass (Tricot–Pellerin *et al.*, 1994).

Symbiotic fixation generally reaches maximal rates when nodule biomass becomes maximal. Afterwards, fixation rates appear to be related to crop growth rate and to carbon competition with reproductive organs as they progressively appear. At this stage, it seems that nitrate availability does not directly affect fixing activity. The peak fixation rate can occur late in the growth cycle, after the beginning of seed filling.

At the end of the growth cycle, symbiotic fixation activity sharply decreases during the seed filling phase. The strong limitation of symbiotic fixation is not only due to competition for carbon with growing seed but also to the age of the nodules, which determines their efficiency. However, this has not been extensively studied and the determination of nodule senescence (nodule age, carbon limitation etc.) remains to be elucidated.

- N_2 fixing activity depends upon nodule biomass

The relationship between symbiotic fixation activity and nodule biomass was studied using $^{15}N_2$ isotopic labelling in controlled conditions on plants, either relying solely on symbiotic fixation or on both N uptake pathways, for their nitrogen nutrition (Voisin *et al.*, 2003). Until flowering, fixation activity was proportional to nodule biomass. Nodule specific activity, as defined by the ratio between symbiotic fixation activity and nodule biomass, does not depend upon nitrate (therefore nitrogen nutrition regime) but it decreases with time. At the seed filling stage, fixation is low and no longer related to nodule biomass.

Before the seed filling stage, as fixation activity is proportional to nodule biomass, it can be defined as the product of "nodule biomass" and "nodule specific activity", which is the level of fixation activity per unit of nodule biomass:

- Analysis of the determinants of changes of fixation activity throughout the growth cycle showed that nodule specific activity depends on the functional efficiency of nodules (gN fixed per g of C respired), that diminishes with nodule age. For a given age of nodules, nodule specific activity is also modulated by the amount of carbon that is allocated to the root parts, following a single relationship whatever the nitrogen nutrition regime.

• Nodule biomass is reduced by the presence of nitrate (Streeter, 1988), which delays nodule establishment. Once nodule establishment has been initiated, their growth and development do not seem to be affected by the presence of nitrate (Voisin *et al.*, 2003).

> To summarize, the adverse effect of nitrate on symbiotic N_2 fixation at the whole plant level could be accounted for by its effect on nodule establishment (delaying their appearance) with no direct effect on their fixation. Indeed, in our experimental conditions, fixation rate per unit nodule biomass and its changes with carbon availability were not modified by the presence of nitrate.

How to optimise and stabilise the nitrogen nutrition of legumes?

Some attempts to increase the plants' ability to fix nitrogen by increasing the number of nodules formed (hypernodulating mutants) or their fixing activity (nitrate tolerant mutants) were associated with depressed shoot growth and yield (Salon *et al.*, 2001). Increasing total N uptake by the crop thus appears to require optimisation in space and time of the complementarities between both N uptake pathways, provided associated carbon costs are taken into account. A better complementarity (or efficient succession of nitrogen nutrition regimes) should make it possible to limit the decrease in exogenous N accumulation during the seed filling period and to optimise remobilisation so that the amount of nitrogen remaining in the straw is as little as possible, and possibly to prolong the duration of seed filling.

Methodologies for measuring symbiotic nitrogen fixation in the field

Christophe Salon, Anne-Sophie Voisin, Olivier Delfosse and Bruno Mary

Summary of measurement methods used in the field to assess the percentage of N derived from atmospheric nitrogen fixation (%Ndfa)

The aim of these measurements is to quantify separately in pea plants:

- Nitrogen arising from symbiotic fixation
- Mineral nitrogen assimilated by roots.

Two main types of method (see box) are available. Only methods based on the ^{15}N enrichment differential between soil and atmospheric N will be described here.

Definition: [15]N enrichment of a soil compartment is defined the proportion of ^{15}N in this compartment in relation to the total amount of N atoms in it ([14]N + [15]N). As a consequence, every compartment has a ^{15}N enrichment value, independently of the labelled fertilizer supplies.

Measurement of products of symbiotic fixation: acetylene reduction assay (ARA), measurement of hydrogen evolution or of ureides in xylem (cf. Chapter: "Carbon and Nitrogen fluxes within the plant" p. 88).
These methods are hard to set up and sometimes not suited (for example ureide measurement) to the study of temperate legumes.
It is not possible to make integrated measurements in the field during the growth cycle. These methods only provide "snap shots" of symbiotic fixation.

Using the difference in ^{15}N enrichment between soil and atmosphere. This difference can be natural or induced by the supply of ^{15}N labelled fertilizer (isotopic methods).
This method allows the integrated symbiotic fixation throughout growth to be calculated
These measurements do not disturb plant growth.

Three isotopic methods for calculating %Ndfa

• Principle of isotopic methods

A fixing plant can accumulate N coming either from the air, whose ^{15}N enrichment is low or the soil, whose ^{15}N enrichment is generally higher. The final plant ^{15}N enrichment thus represents the relative contribution of each of these N sources.

The percentage of N coming from symbiotic N fixation of atmospheric dinitrogen (% Ndfa: *N derived from air*) is calculated from the dilution of the ^{15}N-enriched soil N pool uptaken by the legume plant by the less ^{15}N-enriched atmospheric N pool.

Precaution: for measurements made at the beginning of the growing period, it may be important to take into account a third N source—N arising from the seeds—as this provokes a dilution of the endogenous N.

• Methods using a non-fixing reference plant for the estimate of ^{15}N enrichment of soil mineral N pools

The ^{15}N enrichment of soil mineral N is hard to determine because it varies in time and space. Using a non-fixing plant (a so-called reference plant) is a way of estimating the soil mineral [15]N (with or without fertilizer) that can be accumulated by the legume plant; the N in the reference plant comes entirely from soil mineral nitrogen and so the ^{15}N enrichment of the reference plant is the same as that of the soil mineral nitrogen.

These methods necessitate an appropriate choice of a reference plant having (i) similar shoot and root growth as the legume plant; (ii) access during the growing period to the same soil mineral N pools as the legume plant. The reference plant should be grown at the same site as the legume and receive similar amounts of fertilizer and water. Generally, spring barley is chosen as the reference plant for pea.

Natural Abundance method (NA)

Principle: this method exploits the natural difference in ^{15}N enrichment between soil and the atmosphere (Amarger *et al.*, 1979).

Main hypothesis: [15]N enrichment of the reference plant is assumed to be the same that of the soil mineral N absorbed by the legume plant.

Advantage: this simple method does not involve the use of any labelled fertilizer.

Constraints: its precision is critical when the difference between the ^{15}N enrichment of the soil and that of the atmosphere is less than 0.0015 %, or when the activity of symbiotic N fixation is low (for example at the beginning of growth).

Harvests and measurements

- Measurement of soil N remaining at sowing and its ^{15}N enrichment to see if he latter is sufficiently high.
- For each replicate and measurement date, harvesting a square metre of shoots from both the reference and the legume plant, oven drying at 80°C (48 hours), fine grinding, preparation for determination of total N and of ^{15}N enrichment by mass spectrometry.

Calculation:

$$\%Ndfa = \frac{E_{ref} - E_{leg}}{E_{ref} - \varepsilon_{fix}} * 100$$

% Ndf (Soil) = 100 – % Ndfa

% Ndf (Soil) = N derived from the soil in the legume plant,

E_{ref} and E_{leg} = ^{15}N enrichment of reference and legume plant resp.,

$\varepsilon_{fix} = -1\ \delta$ ‰ for pea.

ε_{fix} is the isotopic fractionation (^{15}N discrimination as related to ^{14}N) that occurs when symbiotic fixation of atmospheric nitrogen takes place. This parameter corresponds to the ^{15}N enrichment of the legume plant when it relies exclusively on symbiotic N fixation. It is measured on plants grown in the greenhouse, without any supply of N.

Isotopic Dilution method (ID)

Principle: this method relies on the addition of ^{15}N labelled fertilizer, which increases the ^{15}N differential between the soil and the atmosphere, and the precision in the estimate of %Ndfa (Rennie and Rennie, 1983). The soil mineral N available for the plants has then two origins: (i) soil N *sensu stricto*: residual N resulting from soil organic matter mineralization and (ii) N arising from the supply of labelled fertilizer.

Main hypothesis: the ^{15}N enrichment of the reference plant (E_{ref}) cultivated with a supply of ^{15}N labelled fertilizer is assumed to represent that of soil and fertiliser mineral N acquired by the legume plant. The hypothesis is that (i) soil is uniformly labelled by fertiliser addition and that (ii) the legume and reference plants take up N in similar proportions from the soil and fertiliser N pools.

Specific treatments: at sowing, a supply of labelled mineral N (ammonium nitrate doubly labelled: ^{15}NO$_3$–^{15}NH$_4$) is sprayed in liquid form at 60 mL.m^{-2} on both the legume and reference plant plots. Each N treatment applied to the legume plant must be also applied to the reference plant. The choice of the ^{15}N enrichment of the fertiliser has to take into account any dilution by N residues.

Advantage: adding labelled fertilizer allows (i) an increase in the precision of the calculation of the amount of symbiotically fixed N (ii) the study of plants having mixed nitrogen nutrition regimes (fixation/assimilation).

Constraints: In most cases, it is tedious to label the soil uniformly by the addition of fertiliser because the soil mineral N and fertiliser N are distributed differently within the soil profile. As a consequence, the hypothesis that the reference and legume plants take up N identically from the soil and fertiliser pools may become unrealistic, due to different root exploration by the cereal reference plant and the legume plant.

Harvests and measurements
- Measurement of N residues at sowing and of their ^{15}N enrichment in order to calculate the ^{15}N enrichment of the fertiliser that has to be applied.
- For each replicate and measurement date, harvesting a square metre of shoots from both the reference and the legume plant, oven drying at 80°C (48 hours), fine grinding, preparation for determination of total N amount and of the ^{15}N enrichment by mass spectrometry.

Calculation:

$$\%Ndfa = \frac{E_{ref} - E_{leg}}{E_{ref} - \varepsilon_{fix}} * 100$$

% Ndf (Soil + Fertiliser) = 100–% Ndfa

% Ndf (Soil + Fertiliser) = N derived from the soil and fertiliser in the legume plant,
E_{ref} and E_{leg} = ^{15}N enrichment of reference and legume plant resp.,
$\varepsilon fix = -1\,\delta\,‰$ for pea.

Method using various ^{15}N enrichment levels: Multiple Enrichment Technique (MET)

Principle: This method uses legume plants which have received fertiliser with different levels of ^{15}N enrichment rather than a « legume and reference plant » pair. It does not require the hypothesis which states that plants from different species have similar soil mineral nitrogen retrieval, both in time and space. The legume plant takes up fertiliser in similar amounts but with various ^{15}N enrichments (Ledgard *et al.*, 1975). A non-fixing reference plant is used only for determining the soil natural ^{15}N enrichment *sensu stricto* (Esol), which varies somewhat with the soil depth.

Main hypothesis: the ^{15}N enrichment of the legume is proportional to that of the applied fertiliser ($E_{fertiliser}$): $E_{leg} = \alpha + \beta \times E_{fertiliser}$

Specific treatments: for each treatment applied to the legume plant (nitrogen fertiliser supply …), at least two other fertiliser treatments, identical in amount but ^{15}N-labelled differently. The reference plant has to be grown in the same conditions as the legume plant, but without any supply of mineral N.

Advantage: Because the hypothesis which states that "the reference plant (usually a cereal) should have similar soil mineral nitrogen uptake as the legume plant" is no longer necessary, the precision is enhanced.

Constraints: each treatment has to be duplicated with at least two ^{15}N enrichment levels in order to be able to estimate with accuracy the coefficients α and β.

Harvests and measurements:

- Measurement of residual N at sowing and of its ^{15}N enrichment in order to calculate the ^{15}N enrichment of the fertiliser that has to be supplied.
- Analysis by mass spectrometry of a sample from the labelled fertiliser solution applied on plants at sowing ($E_{fertiliser}$).
- For each replicate and measurement date, harvest of a square metre of shoots from both the reference and the legume plant, oven drying at 80°C (48 hours), fine grinding, preparation for determination of total N amount and of the ^{15}N enrichment by mass spectrometry.

Calculation: different fertiliser treatments allow the calculation by linear regression of the α and β coefficients using experimental values between E_{leg} and $E_{fertiliser}$; the fertiliser and atmospheric N contributions to the overall N acquisition by the legume plant are calculated as:
Calculation:

$$\%Ndfa =(\alpha +(\beta-1)^*E_{soil})/(\varepsilon_{fix} - E_{soil})$$
$$\%Ndf(fertiliser) = \beta$$
$$\% Ndf(soil) = (\alpha +(\beta-1)^*(\varepsilon_{fix}) /(E_{soil} - \varepsilon_{fix})$$

α and β : linear regression coefficients between E_{leg} and $E_{fertiliser}$
E_{soil}: ^{15}N enrichment of reference plant.
$\varepsilon_{fix} = -1\,\delta$ ‰ for pea.

Conclusions: the MET method has the advantage of allowing the measurement of symbiotic fixation even if fertiliser has been applied, without any need to assume that the legume plant and the reference plant have similar soil mineral N uptake. Nevertheless, a non-fixing reference plant is still necessary to determine the soil mineral N enrichment (E_{soil}).

4

Carbon and nitrogen fluxes within the plant

Christophe Salon, Nathalie Munier-Jolain

The dry matter of pea plant organs usually contains about 40 % of carbon (C) and between 0.4 and 7 % of nitrogen (N). Nitrogen is more concentrated in shoots than in below-ground parts such as roots and nodules; of the shoot organs, the N content of the apices is generally the highest, while that of young leaves or fruits is higher than that of mature leaves, itself higher than the stem N concentration. The partitioning among plant parts of the N and C that has been acquired by the plants from the environment depends upon plant architecture, plant vascular system, biochemical mechanisms involved in the transfer and metabolism of compounds and on interactions between the plant and the environment.

Brief review of C and N transport within plants

The evaporative loss of water from shoot organs drives the upward flow of sap within xylem cells. The flow of photosynthetates and nutrients is driven through the phloem by osmotic pressure differences from source leaves to heterotrophic sinks, where sugars are then consumed either for growth, maintenance or storage.

General trends concerning the compounds loaded into transport pathways

• Sugar and amino acid loading in phloem

Photosynthetically produced glucose is incorporated into sucrose. Only sucrose, some amino compounds, phosphorus and potassium can be loaded into the phloem. Carbohydrates and amino compounds move from their site of synthesis to sieve elements of the phloem either following a cell to cell pathway using small cytoplasmic channels (plasmodesmata) (symplastic pathway), or using extracellular free spaces (apoplastic pathway). Amino compounds may additionally be loaded via the apoplasm from xylem in phloem. Phloem sap compounds subsequently flow toward sinks where they are metabolized: either respired, or incorporated for growth or stored.

• Amino compound loading in xylem

In roots, water and minerals diffuse osmotically via both an apoplastic and a symplastic pathway up to the endodermis cells which surround the conducting tissues (xylem and phloem). Because the cell layer is tangentially suberized (the Casparian strip) at the level of the endodermis, any apoplastic transport is precluded and ions are then actively pumped from the symplasm to xylem apoplast (tracheids).

Transport

• Xylem flow

Constraints and needs
The main driving force for upward movement of solutes in the xylem is suction generated by transpiration. Hence water transport through xylem is apoplastic under tension, which necessitates "pipes" whose structure is able to resist the tension: tracheids and xylem vessel cell walls are composed of lignin (secondary wall) covering the primary cell wall. Root pressure is created by the uptake of water following that of ions but has only a minor role, mainly serving to re-establish continuity in xylem vessels when the rate of transpiration (the main component of xylem flow) is either very high or in some plants reduced due to high air humidity.

Composition of the xylem sap
Xylem sap has an acidic pH (6 to 6.5) and the C/N ratio of its compounds varies from 1.5 to 6. It is composed mainly of nitrogenous compounds

(amino acids, amides,), a small proportion of organic acids and sometimes nitrate ions. Hence, C in xylem sap is mostly that of N compounds while C from organic acids only represents 3 to 30 % of xylem C. The main solutes in xylem sap are highly specific to the species (Layzell *et al.*, 1981): asparagine is the main nitrogen compound in the xylem of temperate legumes. For tropical legumes, ureides predominate in the xylem. The N compound spectrum found in xylem also depends upon the N nutrition regime: temperate legumes supplied with large amounts of nitrate in the nutrient solution will shift from asparagine export to glutamine export from roots to xylem following nitrate assimilation. Similarly, tropical legumes will export amides instead of ureides following nitrate supplementation. Although nitrogenous compounds found in xylem are similar to those in phloem, in the latter they are 5 to 40-fold more concentrated, depending on the site of sap collection (Pate *et al.*, 1975).

• Phloem flow

Source to sink movement
Assimilates, including photosynthate, travel over long distances via phloem sieve tubes which constitute a symplastic network with companion cells, a continuous pipeline to the sink where sucrose and water leave it:

— the sites of sucrose synthesis, i.e. photosynthetically active organs, have a high sucrose concentration. Active transport of sucrose attracts water by osmosis at the sources;

— at the other end, sinks maintain an osmotic gradient by either (i) respiring sucrose, (ii) storing it in vacuoles, or (iii) converting it to osmotically inactive forms (starch, cellulose, proteins etc.).

As such, in contrast to xylem, phloem circulation does not rely on transpiration flow but depends upon the occurrence of an osmotic potential differential between both ends of the xylem vessels : the xylem sucrose flow is symplastic under pressure. This permits circulation of elaborated phloem sap, in particular sucrose, from source organs to sink organs, even when these have a low transpiration rate. As such the functioning of plant organs which do not transpire actively depends generally on assimilate supply via phloem: this is the case for fruits, apices and growing organs.

Phloem sap composition
Phloem sap has an alkaline pH (7 to 8.5) and a C/N ratio between 10 and 200. Phloem transports mainly sugars (mostly sucrose and a very small amount of hexose except in wounded tissues), alcohol sugars and amino acids (aspartate and glutamate and their corresponding amides, asparagine and glutamine). Potassium is the more abundant ion in phloem sap, which never contains nitrates.

C and N fluxes and partition of assimilates

The most documented literature on C fluxes in plants arose from pioneering studies conducted on legumes by Pate (Pate and Layzell, 1981; Pate *et al.,* 1979b). The modelling of the acquired data provides C and N flow charts for the different compartments: amount of photoassimilate produced, amount of fixed or assimilated N, incorporation of C and N within each plant organ, C and N flow in phloem and xylem, transfer from xylem to phloem and from phloem to xylem. These flow charts provide information about:

- mechanisms of translocation and assimilate exchange within plants,
- identification of organic and inorganic solutes moving in xylem and phloem,
- functional C economy of nodules and nodulated roots in relation to their efficiency to convert photosynthates from the shoot into nitrogenous compounds produced by symbiotic N_2 fixation or root assimilation of soil mineral nitrogen,
- the role of vegetative parts in the synthesis and transport of assimilates necessary for feeding fruits and seeds,
- the metabolism of fruits which convert their own photosynthetic products into seed reserves and the translocated compounds received from the mother plant.

C and N phloem and xylem flux

- C/N ratio of sap and plant organs as related to their location within the plant

The C/N ratio of phloem-translocated compounds as they leave the leaves is generally higher than that of those arriving at the sink organs. This is because phloem flow is enriched in N by xylem to phloem transfer at the level of the node, which in legumes involves transfer cells (Layzell *et al.,* 1981, Pate *et al.,* 1975). These exchanges occur mostly in the higher part of the plant and allow N from the transpiration stream to be attracted or recycled into the phloem and then to growing parts (apices, young leaves, fruits) which have a low transpiration demand, allowing these organs to acquire more N than if they were fully dependent on xylem for their N supply. As such, the lowest C/N ratio of phloem sap is found in the shoot apex.

- C transport associated with N compounds; consequences for N flow

Nitrogenous compounds are exported by roots to shoots via the xylem but the transfer from xylem to phloem thereafter varies according to the electrical charge of the compounds. This being so, taking into account the acidic pH of xylem sap:

- amino acids in cationic form such as arginine are fixed on the negatively-charged cell wall bordering xylem vessels (adsorption): these amino acids allow an efficient retrieval by vascular tissues of stems, petioles and major veins of the leaves (Pate *et al.*, 1979a; McNeil *et al.*, 1979);
- neutral forms (asparagine, valine, glutamine, allantoin) are distributed in relation to the capacity and selectivity of membrane uptake by cells bordering xylem in stems. These neutral compounds are more accessible for xylem to phloem transfer.
- anionic forms (allantoic acid, nitrate, aspartate and glutamate) are excluded from both xylem cell wall adsorption and exchanges with xylem protoplast. They are delivered unaltered to transpiring organs such as leaves. Aspartate and glutamate accumulate mostly in non-vascular leaf regions.

The presence in xylem sap of compounds not easily transferred to phloem (arginine, asparagine and glutamine) ensures that nitrogenous compounds arising from root assimilation are retained by mature parts of aerial parts, in particular cell mesophyll, for synthesis of leaf proteins needed for C assimilation (Pate *et al.*, 1981).

Compounds easily exchanged between xylem and phloem (asparagine, glutamine and valine) have a major importance for loading upward flow (mostly asparagine) or downward flow (mostly glutamine) of translocated compounds with N, hence providing N to meristematic tissues during their growth, and to developing fruits (Pate *et al.*, 1981). Amides (asparagine and glutamine) which are the main compounds concerned with xylem to phloem transfer are therefore beneficial for the nitrogen nutrition of such tissues because their low C/N ratio (2 and 2.5 resp.) enriches the phloem in N.

Characteristics of C and N flux according to the organs

- Roots and nodules

Root (and nodulated root) respiration, including the respiratory cost of synthesis, maintenance and metabolism of the below-ground organs, consumes up to 30 % of the C acquired by plants (Layzell *et al.*, 1981) and up to 80 % of C assimilates coming from the shoot and allocated to below-ground parts (Voisin *et al.*, 2003).

The rest of the C transported by shoots to roots via the phloem either provides the carbon skeletons of nitrogenous compounds or returns via the xylem as organic acids.

Hence, due to the extent of root respiration, the dry matter C/N ratio of roots is much lower than that of the flow of translocates feeding them. Root respiration per unit dry weight is typically higher than that of shoots, but part of the respired CO_2 can be re-assimilated via Phosphoenol pyruvate carboxylase. The compounds exported by roots in xylem come not only from soil mineral N assimilation by roots or atmospheric dinitrogen symbiotic fixation within nodules, but also from other catabolic processes, export of previously stored soluble compounds, or recycling from phloem to xylem in the roots. These latter processes however represent only about 15 % of the N exported by the roots.

- ## Stems and petioles

Stems and petioles actively rely on leaf photosynthates for their dry matter production and the maintenance of their respiratory activity.

- ## Apices

Because apices have a low respiration rate, the C/N ratio of their dry matter is very similar to that of the translocate they receive: there is a preferential nitrogen transfer toward apices as compared to roots and 40 % of the flow of N arising from leaves comes from the xylem transpiration stream, the rest coming from phloem (Layzell *et al.*, 1981).

- ## Leaves according to their developmental stage

During the morphogenesis of young developing leaves, C and N are being imported from phloem sap to build up the leaf structure (leaf area) (fig. 1.34; Phase 1).

When leaves begin to transpire, the xylem contribution to their N supply increases up to 80 % of the total N needed for leaf growth, and leaves begin to export C via phloem sap (fig. 1.34; Phase 2).

Apart the very early phase of its development, where they import exclusively C and N by phloem (fig. 1.34; Phase 1), leaves constantly recycle C and N by xylem to phloem transfer (fig. 1.34, Phase 2, 3 and 4).

Compounds from the xylem enter leaves via the apoplast as they transpire and are then redistributed to growing parts of the plant with low transpiration rates. During leaf development, the carbon content of phloem sap decreases because the photosynthetic activity of the leaf is greatly reduced and concomitantly, N content increases because of enhanced mobilisation of N. The photosynthetic activity is then progressively reduced by N remobilisation (fig. 1.34; Phase 3 and 4).

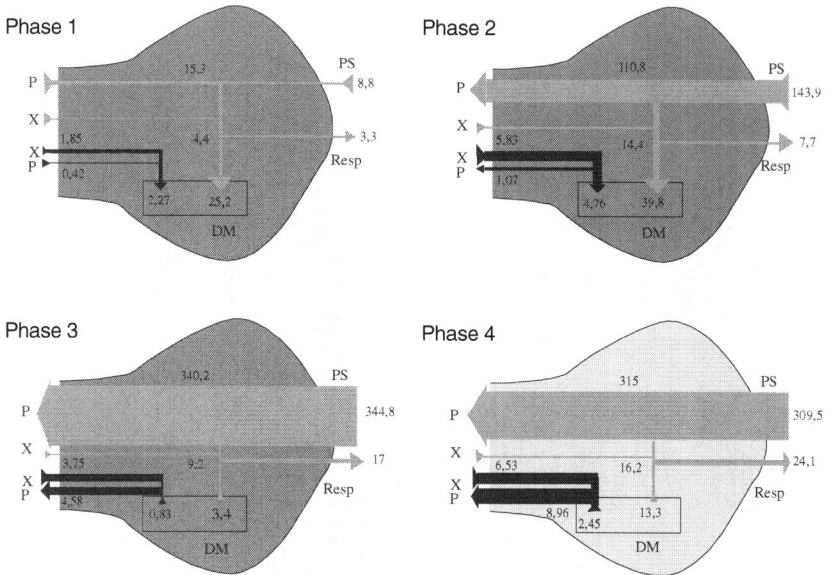

Figure 1.34. Phases which characterize economy of C and N in leaves of the top of the plant in white lupine. The thickness of lines is proportional to C and N fluxes expressed in mg C or N. X =xylen; P =phloen; PS =photosynthesis; Resp = respiration; from late Atkins, 1983—reproduced with the kind authorizatiohn of American Society of Plant Biologists.

• Pods and seeds

The sucrose concentration of phloem sap arriving at reproductive nodes increases with the node number (Munier-Jolain and Salon, 2003), due to xylem to phloem transfer (Pate *et al.*, 1975; Layzell *et al.*, 1981) and/or enhanced contribution of the neighbouring leaves to the C economy of the closest fruits (Pate, 1980). Over the growth cycle, xylem furnishes about 10 % of the fruits' N needs, while the phloem supplies the rest. During fruit development, the sucrose phloem content declines, whereas its amino acid content increases. Two phases can be distinguished (Pate *et al.*, 1977, Peoples *et al.*, 1985):

- • the first one is the expansion of pods and a low seed growth rate, with little C and N accumulation within seeds but intense cell mitotic activity within the embryo (cf. Chapter I.5.2.). During this period net photosynthetic carbon gains by pods equal the losses by respiration. The sucrose flux delivered to the seeds during this period is a key determinant of the extent of cell division within embryos (Munier-Jolain and Salon, 2003);

- the second phase corresponds to a large C and N storage in pods and to active seed growth: the C and N seed needs are then very large. They can be satisfied within pods via (i) mostly phloem sap, (ii) pod photosynthesis. When pods are fully exposed to the light they are near the CO_2 compensation point and the C balance is slightly negative, while fruits lower in the canopy have a respiration much higher than their photosynthesis. Because the internal fruit atmosphere is enriched in CO_2 (10,000 to 15,000 ppm) pods can also photosynthesise CO_2 arising from seed respiration but it depends upon the fruit's position in the canopy : the photosynthetic activity of pods may then improve the fruit C budget up to 20 % (Atkins *et al.*, 1977; Flinn *et al.*, 1977). At the end of this important storage phase, photosynthesis decreases sharply and the fruit loses much carbon via respiration.

Biomass and nitrogen partitioning during crop growth

Marie-Hélène Jeuffroy, Nathalie Munier-Jolain, Jérémie Lecoeur

The sequential production of organs on a pea plant leads to a great variability in the nature, the developmental stage and the number of sinks present on a plant at a given date. From emergence until the end of flowering, which marks the end of appearance of new leaves, vegetative organs are growing. During the same period, the roots and nodules, which are heterotrophic organs, are growing and require assimilates. Between the beginning of flowering and the end of the final stage in seed abortion, flowers and young pods are growing. From the beginning of seed filling until physiological maturity, seeds are filling and accumulate large amounts of carbon and nitrogen reserves. Even in optimal growing conditions in the field, the plant cannot produce the assimilates required for maximal growth of all its organs. Different methods presented in this part can help to explain the rules of carbon and nitrogen assimilate partitioning among the various organs of the plant throughout growth.

The balance-sheet for biomass: dynamic description of biomass and nitrogen accumulation in the various aerial organs throughout growth

The balance sheet method for biomass allows the net influx of assimilates in an organ to be calculated, corresponding to the difference between the input and the output of assimilates for this compartment. For biomass, a net accumulation corresponds to the amount of assimilates translocated through the xylem or the phloem to a compartment (cf. Chapter: "Carbon and Nitrogen fluxes within the plant") minus the amount of carbon respired and the amount of carbon lost by remobilisation to other organs. In the same way, the nitrogen increase in a compartment corresponds to its entry into this compartment minus the amount of nitrogen translocated through other organs and the amount lost by volatilisation through the atmosphere (Schjoerring and Mattsson, 2001).

- ## Biomass allocation to the different vegetative and reproductive organs

The biomass increase in the various organs of a plant is not synchronous (fig. 1.35).

During the vegetative period (I), from emergence until the beginning of flowering, three vegetative compartments have a net biomass increase: the leaves, the stems and the root system. Both the stems and leaves have regular growth, with a biomass allocation constant through these two compartments, between 65 and 75 % of the aerial biomass being allocated to the leaves. This phase corresponds to the period of formation of the structures of CO_2 and radiation capture (cf. first chapter on Vegetative Development). At the end of this period, that is to say at the beginning of flowering, only 25 % of the final total biomass of the plant is generally already accumulated; however in a pea crop, the leaf area already established at this stage allows the interception of 80 % minimum of the incident radiation, and thus a maximal growth of the crop after this stage. Voisin *et al.* (2002) indicate that the growth of the root system continues during all the vegetative period (cf. chapter on root and nodule establishment), but represents only 5 to 10 % of the total biomass at the beginning of flowering (Vessey, 1992; Herdina and Silsbury, 1990; Voisin *et al.*, 2002).

From the beginning of flowering until the beginning of seed filling (II), the biomass accumulation continues in the stems and leaves, while pod growth is slow. During this period, pod growth essentially results from the growth of the pod walls; cell divisions occur in the seeds but without reserve accumulation (Ney *et al.*, 1993). Thus, during the first part of the reproductive period, from the beginning of flowering until the beginning of seed filling, the vegetative compartments constitute large sinks for assimilate allocation.

From the beginning of seed filling (III and IV), the growth of the leaves and stems slows down greatly, and even stops, while the seeds begin to accumulate large amounts of reserves. Filling seeds have priority in assimilate allocation to the various sinks (Munier-Jolain *et al.*, 1998; Chapter

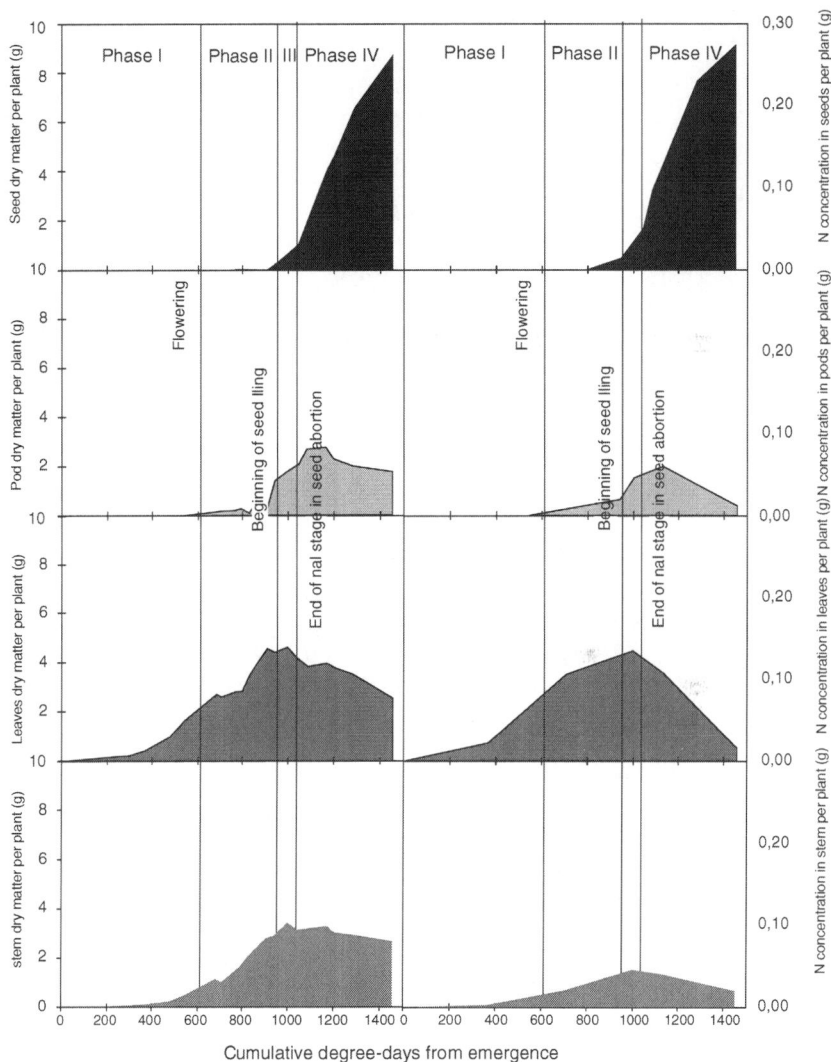

Figure 1.35.

on individual seed weight). Thus, their presence on the plant can lead to the end of emission of new phytomers, because of assimilate competition between sinks (Turc and Lecoeur, 1997; cf. part of vegetative development). Regarding pods, the growth of the pod wall occurs before the period of assimilate accumulation in the seeds (Wang and Hedley, 1993). Thus, when all seeds are filling, that is to say after the end of the final stage of seed abortion, the growth of the pod walls is finished. From the end of the final stage of seed abortion until physiological maturity, the seeds are the only sink still accumulating biomass, the growth of the other sinks having stopped; yet carbon is allocated to these other organs for their respiration. During the entire period of seed filling, the amount of biomass in the leaves and in the pod walls slightly decreases: this decrease can be due to (i) a loss of biomass from the leaves (leaf fall, senescence), (ii) a loss of CO_2 through their respiration, when it is higher than the gross photosynthesis, and (iii) a remobilisation of carbon and nitrogen to the seeds.

- ## Carbon and nitrogen partitioning during growth

The carbon content, expressed as a percentage of the biomass of the various compartments, does not change a lot over time for a given organ or between organs : the average carbon concentrations during growth for the stems, the leaves, the pod walls and the seeds are 43.13 % (± 0.27), 43.34 % (± 0.59), 42.65 % (± 0.45), 42.76 % (± 0.77) respectively. These values are close to those observed in other grain legumes (Pate *et al.*, 1983). The fairly uniform carbon concentration, between 42.5 % and 43.5 %, of the leaves, stems and pods, indicates a very low remobilisation of carbon in the plant (Warembourg *et al.*, 1984). The carbon assimilated by the photosynthetically active leaves is directly distributed to the sink organs for their metabolic activity and/or growth. Thus 80 % of seed filling is attributable to photosynthetic activity during the filling period (Pate *et al.*, 1983 sur *Vigna unguiculata*) and a low temporary storage is observed in the pea.

The nitrogen concentration of the organs varies greatly during growth (fig. 1.36) : the nitrogen concentration of all the organs is very high during the period of cell division (> 7 %). After this phase the nitrogen concentration of the organ decreases.

From emergence until the beginning of seed filling, the amount of nitrogen increases in all the vegetative organs (stems and leaves) and in the pod walls. For all these compartments, this nitrogen accumulation is always associated with accumulation of biomass (fig. 1.35). This net influx of nitrogen in these organs is still slower than biomass accumulation, which leads to a nitrogen dilution in the organs and in the whole plant (cf. p. 61). For a leaf, the amount of nitrogen decreases from the stage of leaf maturity. At this stage, the leaf area no longer increases, but its nitrogen concentration (%) decreases (fig. 1.36). Consequently the extra nitrogen accumulated in the leaves is not equally distributed among the leaves according to their age : the amount of nitrogen in the mature oldest leaves decreases, this nitrogen being translocated to the best illuminated levels of the crop in order to maximise photosynthesis (Hirose and Werger, 1987).

From the end of the final stage of seed abortion, the amount of nitrogen decreases in all compartments (stems, leaves, pod walls) except the seeds.

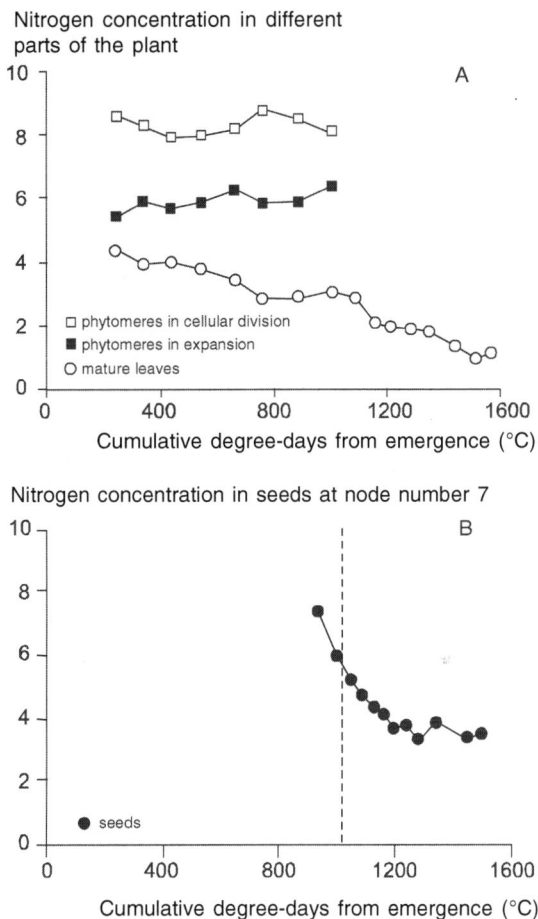

Nitrogen concentration in different parts of the plant

□ phytomeres in cellular division
■ phytomeres in expansion
O mature leaves

Cumulative degree-days from emergence (°C)

Nitrogen concentration in seeds at node number 7

● seeds

Cumulative degree-days from emergence (°C)

Figure 1.36.

This is because the N requirements for the synthesis of protein reserves in the seeds is very expensive: the accumulation of exogenous N during seed filling, either from N_2 fixation, or from the absorption of mineral N, only represents 25 to 40 % of the total N entering the seeds (Lhuillier-Soundélé, 1999; Crozat *et al.*, 1994). Thus a large part of the N accumulated before the beginning of seed filling is remobilised to the seeds (Sinclair and de Wit, 1976, Peoples *et al.*, 1983b, Lecoeur and Sinclair, 2001b), which is not the case for carbon. The leaves represent the largest source of remobilised N to the seeds: Peoples *et al.* indicate that the leaves, roots and stems contribute to the N remobilisation to the seeds in the proportions of 5:2:1 respectively.

Rules of carbon assimilate partitioning among the organs during the reproductive phase

Although the literature shows that some preferential relationships exist between the source leaf and the pods at the same node (Flinn and Pate, 1970), translocation of assimilates occurs between the various nodes of the stem, for example in the case of the reduction of assimilate production by one leaf (Jeuffroy and Warembourg, 1991) or of a low requirement from the sink (Lovell and Lovell, 1970).

By plant labelling with ^{14}C, Jeuffroy and Warembourg (1991) showed that the growing apex has, just after the beginning of flowering, a high specific relative activity (ratio of the proportion of radioactivity in one organ to the proportion of its carbon content) in comparison with the other organs of the plant. These organs are thus strong sinks at this time, whereas their capacity to attract assimilates decreases steadily later as the end of flowering approaches.

The allocation of carbon assimilates to the nodules decreases as soon as pod growth becomes rapid, explaining the reduction of N_2 fixation generally observed in the field from the beginning of seed filling (Lawrie and Wheeler, 1974; Silsbury, 1990).

By pod ablations on pea plants, Hole and Scott (1983) demonstrated competition for the assimilates available on the plant among the different organs. As pods maintain a constant growth rate after the final stage of seed abortion if plant biomass production declines (Munier-Jolain et al., 1998), they take priority in assimilate partitioning over the other sinks on the plant (vegetative organs and young pods). Conversely, among the pods before the final stage of seed abortion, and for which seed abortion may still occur, a hierarchy between sinks in the allocation of assimilates does not seem to exist. The amount of ^{14}C allocated to each pod depends on its initial biomass (Jeuffroy and Warembourg, 1991).

These rules for C partitioning among organs, between the beginning of flowering and the end of the final stage of seed abortion, have been tested in field conditions varying in the date and density of sowing and the year (Jeuffroy and Devienne, 1995). The simulated values of the amount of biomass allocated to the vegetative organs and to the individual pods are very close to the observations, thus validating the rules of assimilate partitioning under agronomic conditions.

Harvest index for biomass (HI) and nitrogen (NHI)

In numerous species such as maize, sorghum, peanut, wheat, barley and sunflower, the biomass partitioning between vegetative and reproductive organs during the final stages of growth can be described with the harvest index for biomass (Bindi *et al.*, 1999). This variable represents the fraction of the total biomass represented by the seeds. It increases linearly from the beginning of seed filling according to cumulative degree-days. The rate of HI increase is generally constant during seed filling for a wide range of conditions and is a characteristic of a species or a group of genotypes. The final value of the harvest index is generally constant or inversely correlated with the final value of the total biomass. From these results, in numerous species, biomass partitioning towards the end of growth is simply described by this variable (Bindi *et al.*, 1999).

• Harvest index measured at harvest

In peas, the harvest index is very variable. Its value ranges between 0,30 and 0,65 and is not correlated to the total biomass of the plant (fig. 1.37) (Lecoeur and Sinclair, 2001a). On the other hand, the nitrogen harvest index, corresponding to the fraction of total plant nitrogen recovered in the seeds, is very constant for a wide range of yields (Lecoeur and Sinclair, 2001b). A mean value of 0,80 was determined for a range of conditions including water and heat stress and nitrogen deficiency, corresponding to yields varying between 36 and 613 g.m^{-2} (fig. 1.37). The 20 % of nitrogen still in the vegetative organs at harvest corresponds to structural nitrogen which cannot be remobilised. Conversely, biotic stress often leads to a reduction of the remobilisation of N from vegetative organs, thus increasing NHI.

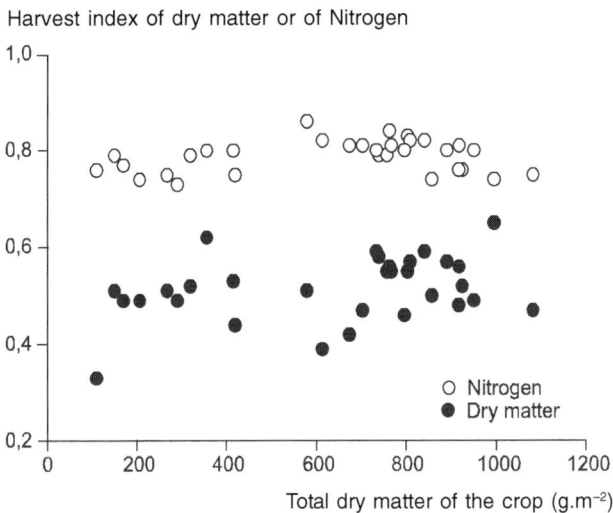

Figure 1.37.

• Harvest index measured during seed filling

Between the beginning of seed filling and physiological maturity, the harvest indices for biomass and nitrogen increase linearly with time expressed in degree-days (fig. 1.38) (Lecoeur and Sinclair, 2001a and 2001b). Nevertheless, the rates of increase of the indices are highly variable among situations. The HI for biomass can vary twofold, with a mean of 1,18 mg.g^{-1}.°Cd^{-1}. The NHI has the same variability, with a mean of 1,65 mg.g^{-1}.°Cd^{-1}. These variations do not depend on temperature or on the rate of biomass accumulation during seed filling. Although the rates of

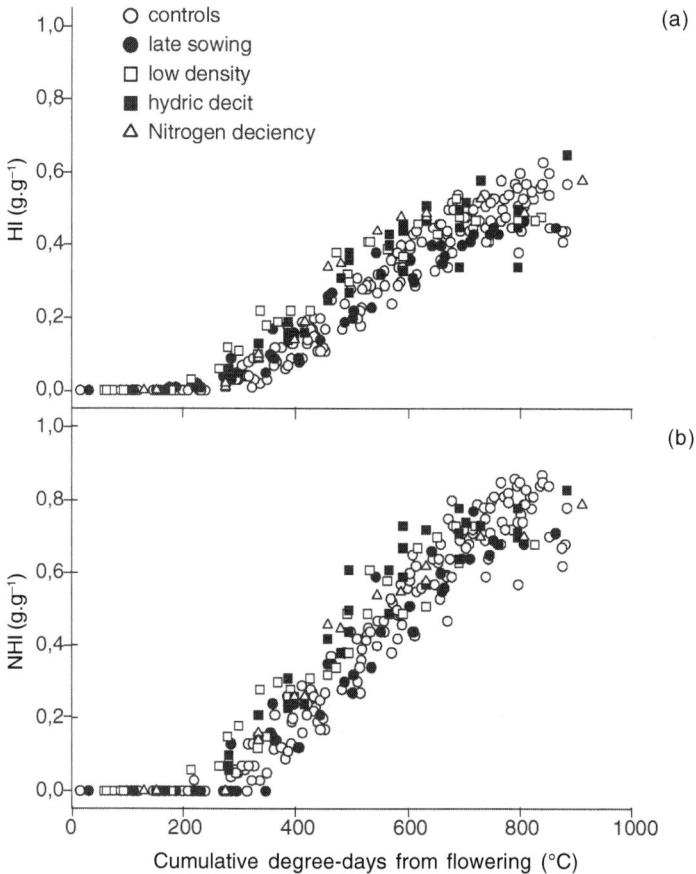

Figure 1.38.

biomass and nitrogen remobilization vary during seed filling, the time-course change of HI and NHi is linear when plotted against time expressed in degree-days (Lecoeur and Sinclair, 2001b and 2001c). In the case of NHI, the linearity is linked with the apparent priority of N remobilisation for biomass accumulation and remobilisation during seed filling. The final period of crop growth is conditioned by the time-course of several variables. The stability of the final level of nitrogen remobilization from aerial parts, in comparison with the biomass remobilization, shows that the maturity date is determined in most cases by this level of nitrogen remobilization (cf. p. 115). The nitrogen harvest index seems to be a suitable variable to characterize the nitrogen management within the plant at the end of growth. The analysis of the simultaneous evolution of the biomass and nitrogen accumulation and of grain development indicates that high yields (above 700 g.m^{-2}) can only be obtained if nitrate assimilation is maintained at a rather high level at the end of growth, since nitrogen fixation generally decreases just after the beginning of seed filling. Another condition to maximise yield and harvest index is to satisfy the seed biomass requirements until physiological maturity. As individual seed growth dynamics vary little, the biomass requirements of the reproductive sink depend on the seed number. This is linked with the growth rate of the plant during the 10–15 days following the beginning of seed filling, whatever the level of biomass accumulation and the constraints occurring during this period (Guilioni *et al.*, 2003). The indeterminate habit of the pea plant means that the period of determination of seed number may vary in length. This characteristic gives the pea some plasticity in response to constraints which may occur at the beginning of the reproductive development.

These results suggest that yield and harvest index are determined by three variables: (i) the amount of total biomass accumulated which, multiplied by the harvest index, gives the potential yield, (ii) the total amount of nitrogen accumulated, which determines a maximal lifespan and hence the biomass accumulation in the seeds and (iii) the growth rate of the plant between flowering and the final stage of seed abortion, which determines the accumulation rate of biomass in the seeds through their assimilate requirements. These three variables follow partially independent time courses during reproductive development. They can be affected by abiotic and biotic constraints during this phase. These sources of variation affect the yield and the harvest index and could explain their high variability observed in peas. Unlike in other species, biomass remobilisation to the seeds does not determine yield. The final value of the harvest index is only the consequence of the simultaneous evolution of the biomass accumulation, the nitrogen remobilization and the elaboration of the seed number.

5

The seed number

Marie-Hélène Jeuffroy, Jérémie Lecoeur, Romain Roche

The pea yield, highly variable among fields and years, is highly correlated to the seed number per m² (Hardwick, 1988; Doré *et al.*, 1998). At the plant scale, the period of seed formation begins at the anthesis of the first flower and ends when the last pod reaches the final stage of seed abortion, corresponding to the beginning of seed filling of this reproductive node (Ney and Turc, 1993; cf. fig. 1.10). The indeterminate habit of the pea induces a sequential development along the stem: thus on a same plant, the seeds are simultaneously formed on several reproductive nodes. The total number of reproductive nodes is highly variable among varieties, but also among fields in a same year (Heath and Hebblethwaite, 1987; Turc, 1988; Roche *et al.*, 1998), whatever the presence or not of water and thermal stress (Lecoeur, 1994; Guilioni *et al.*, 2003; Sagan *et al.*, 1993) : the number of reproductive nodes can then limit the final seed number and the yield (Pate, 1975; Murfet, 1982; Jeuffroy, 1991). Moreover, the seed partitioning on the several nodes of a stem can vary not only with the number of reproductive nodes, but also among situations with a same number of nodes but differing in variety, the sowing date or density or the environmental constraints (Turc, 1988; Jeuffroy and Devienne, 1995; Guilioni *et al.*, 2003). The vertical partitioning of the seeds along a pea stem can be a diagnosis tool (Jeuffroy, 1994, cf. p. 230) or a tool to predict the effects of environmental factors on final yield.

Determinants of the seed number at the crop scale

On other numerous species, a highly stable relationship exists between the seed number per m² and the biomass accumulated in the crop during the period of seed formation (wheat: Boiffin *et al.,* 1981; maize: Hawkins and Cooper, 1981; soja: Jiang and Egli, 1995; oilseed rape: Leterme, 1985). In pea, the seed number is fixed between the beginning of flowering and the final stage in seed abortion of the last pods (Pigeaire *et al.,* 1986). It thus seems interesting to analyse the seed number variability relatively to the biomass accumulation during this phase.

A linear relationship between the seed number per m² and the crop growth rate during the period "beginning of flowering—FSSA" was established from a large database covering various varieties, in greenhouse or in fields, without water and thermal stress or mineral shortage (table 1.3; cf. fig. 2.3 p. 145). This relationship is unique and stable in this large range of situations (Guilioni *et al.,* 2003). The stability of this relationship is similar if it was established at the individual plant level or at the crop level, reinforcing the interest of taking the stem as the mean organ in the analysis of the crop functioning.

The formalism is highly simple as it is a proportional relationship between the crop growth rate during the period « beginning of flowering—FSSA » and

Table 1.3. SNCGR : Ratio of the total seed number per m² (SN) and crop growth rate (CGR) during the phasis of seed number formation, from beginning of flowering (BF) to final stage in seed abortion (FFSA). Experimental situations without water or heat constraints and without mineral deficiency. (after Lecoeur et Sinclair, 2001 and Guilioni *et al.,* 2003).

Factor of variation	SNCGR(g °Cj⁻¹)	Mean Thousand Seed Weight (mg)
Cultivar Aladin	1902 a	295
Alex	1920 a	280
Atol	1846 a	230
Baccara	1843 a	290
Frilène	1808 a	200
Messire	1811 a	270
Solara	1803 a	300
Density (pl/m²)		
10	1848 a	
20	1874 a	
40	1805 a	
80	1839 a	
Crop conditions		
Glasshouse	1821 a	
Field	1844 a	

Values followed by letter a are not significantly different at the threshold of 1 per 1000 (P=0,001)

the final seed number. When the plant growth rate is expressed in grams per degree-day, the mean value of the coefficient of proportionality is 1835 ± 75 seeds on the range of cultivars tested ($r^2=0.914$ and RMSE=206 on the whole situations). This study confirms the results obtained previously on the cv Solara, with a coefficient of 1866 (Ney, 1994). No significant difference occurred between varieties in the range of varieties tested, varying in architecture and mean weight per seed. This relationship does not vary according to the crop density, in a large range from 10 to 80 plants.m^{-2}, but also to the growing conditions, particularly the controlled conditions in greenhouse. The variability of these conditions, particularly the lower radiation in greenhouse, is probably already taken into account in the variable "accumulated biomass in the plant". The stability of this relationship allows to use it as an interesting tool for the prediction of yield and its component of a pea crop, when it is possible to estimate the biomass production from climatic data and from an estimation of the intercepted radiation by the crop (cf. Chapter: "Carbon acquisition at the crop level in pea").

However, in this aim, it is necessary to precisely know the period of grain formation. The two stages at the beginning and end of this period are not so easy to determine each others. The beginning of flowering is easy to record, generally measured in all the experiments and diagnosis, and can be easily simulated from climatic data (cf. p. 19; Roche et al., 1999). On the contrary, the end of final stage in seed abortion on the stem requires to follow the evolution of the seed size, of the pod diameter or of the water concentration of the seeds on each reproductive nod, which is time-consuming (cf. p.25), as the present models do not allow to simulate this stage from data easily accessible. A sensitivity analysis showed that it was possible to replace the whole period of seed formation by a shorter period, during 400 degree-days from the beginning of flowering, for the estimation of the rate of biomass accumulation by the plant. This change simplifies the approach and do not lead to errors in the estimation of the final seed number higher than 10 %. It could be possible that this simplification do not apply when abiotic and biotic stresses occur during the period of seed formation. In these cases, the period of seed formation which must be taken into account to calculate the plant growth rate could be different. For a large range of hydric and thermic stresses, it was shown that the relationship calculated between the beginning of flowering and the FSSA is not modified (cf. Chapters: "Influence of water deficit on pea canopy functioning" and "Effects of high temperatures on a pea crop", Guilioni et al., 2003). Finally, it would be necessary to test the relationship for varieties with a mean weight per seed different from those tested, particularly in the case of winter peas, for which mean weights per seed are often low (< 200 mg).

The time-course change of assimilate partitioning: determinant of the seed number profiles

As it was shown just before, the seed number of a pea crop is highly correlated to the crop growth during the period of their formation. At the pod level, the period of seed formation begins at the anthesis of the flower and ends at its final stage in seed abortion (cf. fig. 1.10; Pigeaire *et al.*, 1986). During this period, the pod length increases, but the pod is always flat. An exponential relationship between the pod biomass and its length was shown on pods sampled at various stages between anthesis and FSSA of the seeds inside (Jeuffroy and Chabanet, 1994). Therefore, the measurement of the pod length allows to follow its biomass in a non destructive way. In various conditions of carbon nutrition, the variability of seed number among pods is explained by the variations of the pod length rate during the first half of the period « anthesis—FFSA » (Jeuffroy and Chabanet, 1994).

The individual pod growth depends on the amount of assimilates which is allocated to it and, consequently, on the production and partitioning of assimilates among the different sink organs of the plant. The pattern of assimilate partitioning, presented in the chapters "C and N fluxes within the plant" and "Biomass and N partitioning during crop growth", allow to estimate the amount of assimilates allocated to each pod according to its biomass and on the stages of the other organs present on the plant simultaneously and according to the total amount of assimilates produced by the plant (Jeuffroy and Devienne, 1995). The integration of these processes (dynamics of development and growth of the aerial organs) inside a crop model allows to estimate the seed distribution among the various reproductive nodes according to the plant growth and according to the temperature conditions (Jeuffroy, 1994). The evaluation of this model, in satisfactory water and nitrogen conditions, shows that the seed profiles simulated are close from the observed ones. However, this model was assessed on the cultivar Solara and its adaptation to other cultivars requires the estimation of some specific parameters varying with varieties, some of which being difficult to measure (Gouzonnat, 1992). One interest of this model is to give a potential seed profile, useful for an agronomic diagnosis in farmers' fields (Doré, 1992; Jeuffroy, 1994). Yet, its use is not largely spread because of the estimation of the cultivar parameters and of some variables necessary to initialise the model.

A static model of the seed distribution among reproductive nodes

The dynamic description of the seed profiles formation leads to describe precisely the assimilate partitioning during the reproductive phase. When the modelling of this process at a fine scale is not required, a static model of the final seed profile on the stem can be proposed. This type of model, although less detailed, is often more robust than a detailed approach, because of its simplicity.

Very few studies concern the description of the seed profiles among nodes, but this mathematical problem is close to the analysis of the distribution of the leaf area among nodes. Dwyer and Stewart (1986) and Dwyer *et al.* (1992) on maize, Muchow and Carberry (1990) and Carberry *et al.* (1993) on sorghum and Bouchard (1997) on wheat used the same mathematical equation, simple and requiring few parameters, in order to simulate the area of each leaf, according to the number and the size of the largest leaf. The formalism proposed can be applied to the seed number:

$$SN_i = SN_{max}.exp\ [a.(NN_i - NN_{max})^2 + b.(NN_i - NN_{max})^3] \qquad Eq.\ 1$$

with SN_i the seed number on reproductive node i, NN_i the number of the reproductive node, NN_{max} the number of the node bearing the highest seed number on the stem, SN_{max} the maximum seed number on this node, and a and b are empirical parameters.

The fits of this function to observed data give a good account of the seed profiles (r^2 varying between 0.928 and 1.000) in a large range of situations (Roche and Jeuffroy, 2000). However, the explicative variables NN_{max} and SN_{max} and the parameters a and b vary among situations. For a predictive use of this formalism, it is possible to estimate these 4 variables and parameters from other variables that can be either measured before the seed formation, or simulated by a classical crop model. Thus, NN_{max}, SNmax, a and b depend on the stem number per m² (STNi), on the total number of reproductive nodes (TRN), on the crop growth rate during the period of seed formation (Crop Growth Rate (CGR) in g.m⁻².degree-day⁻¹) and on the total number of seeds per stem (SNT). The equations have been fitted on a sample of situations varying by the year and the location, the date and density of sowing, and corresponding to the only cultivar Solara (Roche and Jeuffroy, 2000):

$$NN_{max} = 2.153 - 0.02437.NTi + 0.004389.NNT.Nti \qquad Eq.\ 2$$
$$r^2 = 0.393,\ rMSEP=0.901$$

$$SNmax = 5.854 + 0.3060\ SNT - 0.2203\ CGR - 0.6830\ TRN \qquad Eq.\ 3$$
$$r^2 = 0.860,\ rMSEP=0.547$$

$$a = -0.2943 + 0.02055 \text{ TRN} - 0.007480 \text{ NGT} + 0.01782 \text{ CGR} \quad \text{Eq. 4}$$

$$b = 0.003832 \text{ NNmax} - 0.003198 \text{ SNmax} \quad \text{Eq. 5}$$
$$r^2 = 0.717$$

The evaluation of these models, realised on a totally independant sample, is satisfactory (see the values of rMSEP, root squared of mean squared error of prediction indicated above), as is shown on the comparison between simulated and observed seed profiles on 3 situations of Solara (fig. 1.39a) and 3 situations of other cultivars (fig. 1.39b). We can see that the diversity of the profile forms are well simulated by the model. On the all situations, the predictive error at the node scale is lower than one seed (Roche and Jeuffroy, 2000).

Finally, the model proposed by Dwyer and Stewart (1986) to simulate the leaf area per node appears as a good tool to simulate the seed number profiles on pea.

Summary

Different methods are available to simulate the seed number per m^2:

- A linear relationship between the seed number per m^2 and the crop growth rate during the period « beginning of flowering–final stage in seed abortion »: the slope of the relationship is 1835 ± 75 seeds in the range of varieties tested. This relationship is valid also in conditions of water and thermal stress;
- A static model rather simple describing the seed number distribution among the reproductive nodes of the stem.

Individual seed weight

Nathalie Munier-Jolain

As for many species, the final yield in peas is highly correlated with variations in the seed number per unit area (cf. previous part on seed number). However understanding the yield formation throughout the growth cycle requires the analysis of individual seed weight:

- The indeterminate growth habit of legumes is characterized by a significant period during which seed set and seed filling occur simultaneously on the plant. During this period young pods bearing seeds still capable of aborting compete for assimilates with seeds that are accumulating reserves in their cotyledons. Consequently the establishment of seed number per node along the stem called the

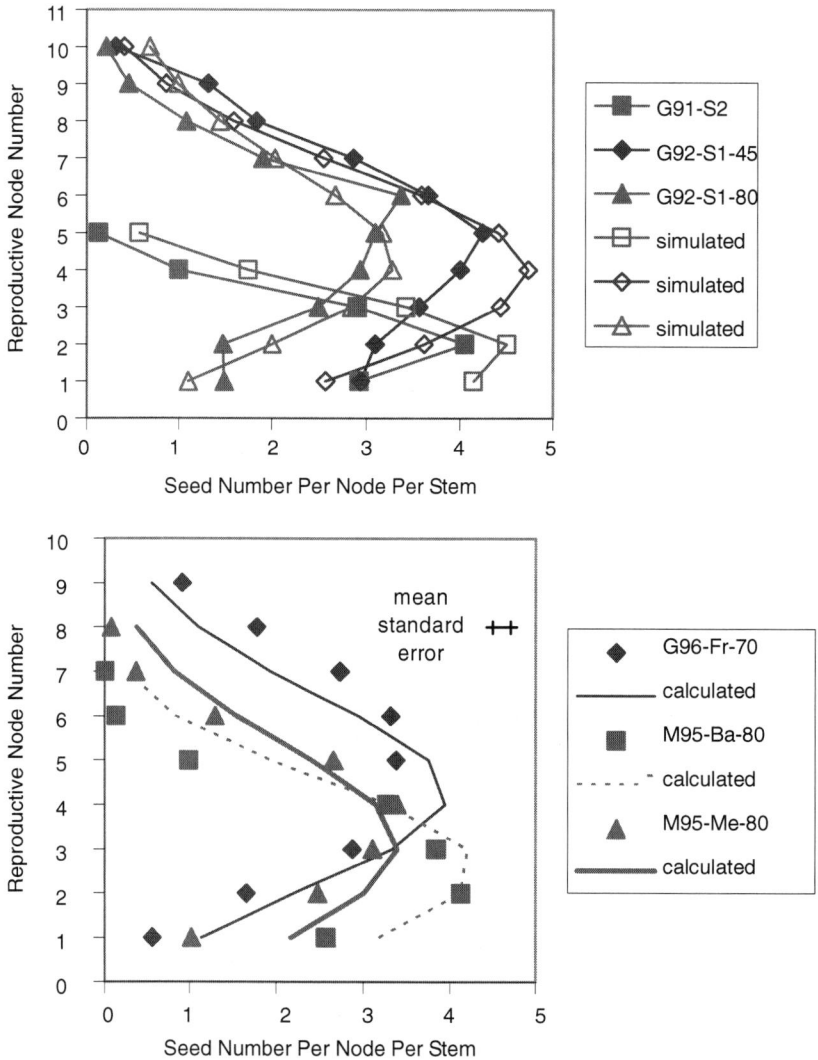

Figure 1.39. Comparison of seed number profiles observed and simulated (from Roche and Jeuffroy, 2000. Reprint with the authorization of the American Society of Agronomy ©ASA) on 3 situations for cv. Solara (a) and on 3 situations for other genotypes (Fr = Frisson, Ba = Baccara, Me = Messire) (b).

"seed number profile" depends on the demand for assimilates of growing seeds already developed at the lower nodes;

- The variability of individual seed weight can still account for the residual variability of yield that remains unexplained by the variations of seed number (approximately 20 %—Source: Arvalis-Institut du Végétal);
- Seed protein concentration is the most important quality characteristic for the sale of grain legumes, and depends on the differential accumulation of carbon and nitrogen during the seed filling period.

Determination of individual seed growth rate

For species with an indeterminate growth habit, the conclusions of different reports on the effect of variations in assimilate availability during the filling period on seed growth remain contradictory (Meckel *et al.*, 1984; Egli *et al.*, 1989). In those species, the indeterminate growth habit results in a sequential pattern for their reproductive development, so that seeds on the same plant achieve different reproductive stages at the same time. Therefore the effects of a change in assimilate availability during the filling period cannot be analysed at the whole plant scale but should account for the development stage achieved by seeds at each morphological position.

• Seed growth rate and cotyledon cell number

Changes in photoassimilates supply during the seed filling period of a given node do not affect the growth rate of its seeds (Munier-Jolain *et al.*, 1998). This indicates that seeds are a priority sink for assimilate as compared with vegetative growth or the formation of new seeds in the upper part of the plant. Thus assimilate flux in growing seeds fluctuates little and we can assume that seed growth rate remains constant during the seed filling period. Consequently seed growth rate observed at the beginning of seed filling corresponds to potential seed growth, which should be determined before the period of rapid seed growth. During the period preceding seed filling, cell division occurs in the embryos and at the end of this period cotyledon cell number is established when linear dry matter accumulation begins.

A relationship between cotyledon cell number, fixed at the beginning of seed filling, and seed growth rate has been established on pea and soybean (Munier-Jolain and Ney, 1998). Consequently seed growth rate observed during the seed filling period is determined before this period, during cell division in the embryo (fig. 1.40).

For the validity domain in which experiments have been conducted (farming practices) the relationship between cotyledon cell number and seed growth rate is valid across genotypes characterised by different seed size. Seed weight differences among genotypes are mainly due to differences in seed cell number, suggesting that the seed cell number is a major factor controlling the seed growth rate and the final seed weight. Further, for a given genotype, differences in seed growth rate among morphological positions on the same plant and in differing environmental conditions, are strongly and positively correlated with differences in seed cell number; seeds at the top of the plant have a higher cotyledon cell number than older seeds located at the base of the plant.

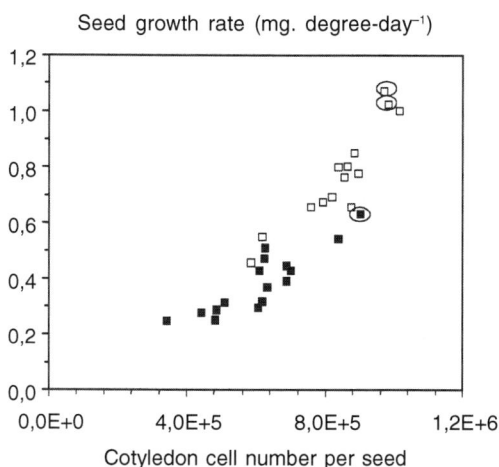

Figure 1.40. Relationship between the seed growth rate and the cotyledon cell number. Pea:Solara, high seeded genotype (open square), Frisson, small seeded genotype (full square). Surrounded symbols correspond to situations where assimilates are non limiting (from Munier-Jolain and Ney, 1998).

• Variability in seed cell number

Seed growth rate is highly correlated to the number of cotyledonary cells produced during the cell division period. As the progression rate of flowering is lower than the progression rate of the beginning of seed filling (cf. fig. 1.10 and p. 26), the duration of the cell division period of upper seeds is shorter than that of seeds produced earlier. Hence, although the duration of cell division is shorter for upper seeds, these have a higher cotyledon cell number than those of the first reproductive nodes. This implies that mitotic activity increases with the position of the reproductive node. To date, several physiological explanations of the variations of mitotic activity in seeds as a function of the nodal position have been cited: (i) partially the C and N supply provided by the plant to the embryos during cell division; (ii) environmental factors such as temperature.

Sucrose flux in phloem sap and seed growth potential

The sucrose concentration in phloem sap supplied to each seed at the upper nodes was higher than that at lower nodes, and higher for the N-stressed plants than for plants having an optimal N nutrition. As seed growth rate is highly correlated with the number of cotyledonary cells produced during the cell division period, individual seed growth rate might be considered as a convenient indicator of cell division between the beginning of flowering and the beginning of seed filling. The relationship between seed growth rate and sucrose flux per seed conforms to Michaelis-Menten kinetics (fig. 1.41).

Beyond any mathematical consideration these results already suggest that the flow of sucrose delivered to the seeds during the cell division period is an important determinant of seed growth potential. Hence it is tempting to speculate that sucrose may be instrumental in the modulation of seed embryo mitotic activity. Knowledge of the plant cell cycle regulation has been gathered and provides a mechanistic explanation. Sucrose modulates cyclin expression which is involved in the regulation of the cell division cycle (Den Boer and Murray, 2000). Moreover the cotyledon cell number at the end of the cell division period is correlated to the hexose concentration in the apoplasm and to the activity of an acid invertase which is involved in the cleavage of sucrose into hexoses (Lemontey 1999).

Apart from sucrose, a low C/N ratio has also been suggested as an important determinant of cell division in meristems. The C/N ratio of the flow of translocates towards seeds was much larger for the N deficient treatment, which displayed the higher seed growth potential. Further the C/N ratio can be low with a high sucrose concentration in the phloem sap, especially for the seeds at the top of the canopy. These results indicate that N flux towards seeds, in itself, is not the main determinant of seed growth potential.

Temperature and cell division

The response of cell division to the mean daily temperature (expressed in number of cells per day) conforms to a bell-shaped curve with an optimal temperature between 18 and 22°C. This relationship between cell division and temperature has been established with seeds cultured in vitro, but is confirmed by estimated cell division rates obtained in field conditions for which temperature has remained stable during the cell division period.

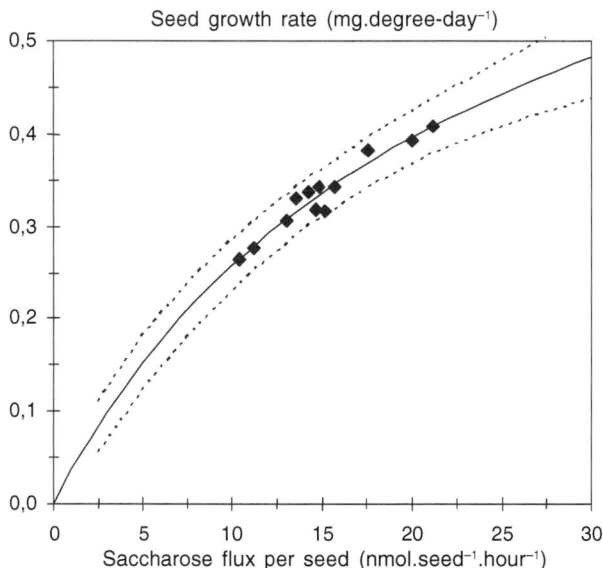

Figure 1.41. Relationship between seed growth rate during its filling and the flux of saccharose per seed during the time of cell divisions (from Munier-Jolain and Salon, 2003).

- ## Variability of the duration of the cell division period in the seed

Little literature is available on the duration of the seed cell division period. For a given genotype in optimal growing conditions, this duration is likely to be constant when expressed in degree-days (Ney and Turc, 1993). However this optimal duration is under genotypic control (Dumoulin et al., 1994). The duration of the cell division period in the seed seems to be higher for large-seeded genotypes and could be under the control of the embryo genome (Lemontey et al., 2000)

Moreover data obtained in field grown plants affected by water, thermic, or nitrogen stresses, indicates that for a given node the duration between flowering and the beginning of seed filling, which corresponds to the cell division period, could be reduced under stress conditions (Sagan et al., 1993; Ney et al., 1994).

Determination of the duration of seed filling

The accumulation of reserve compounds during the seed filling period depends on their photoassimilate demand, determined by their cotyledon cell number. The fulfillment of this demand depends both on the instantaneous carbon and nitrogen assimilation, and on the nitrogen remobilization accumulated in the plant before the seed filling period. Consequently the duration of seed filling at a given node is likely to be affected by assimilate availability (Munier-Jolain *et al.*, 1996; Munier-Jolain *et al.*, 1998). Two mechanisms could lead to the termination of seed filling, depending on the source-sink ratio during the filling period:

- In most agricultural situations an important self-destructive process caused by nitrogen remobilization to filling seed ends seed filling; the high nitrogen demand of the seeds during seed filling triggers the remobilization of proteins involved in photosynthesis and then the production of photoassimilates needed for seed growth ends;
- However in few cases seeds reach their maturity even though nitrogen available for remobilization to filling seeds is not exhausted. In those cases where self-destructive processes are less intensive, carbohydrates could be still synthesized in leaves and thus termination of seed filling occurs because seeds have reached their maximal seed size, which depends on cotyledon cell number and maximal cell size.

• Intrinsic limitation: maximal seed weight

Because the determination of the cotyledonary cell number is presented above, this analysis of the intrinsic limitation of seed size deals only with the variability of the maximal cotyledonary cell volume.

Genotypic effects

As for many species, the endoreduplication process(whereby chromosomes duplicate without any division of the nuclear or the cell) occurs in cotyledons after the beginning of seed filling in pea (Wang and Hedley, 1993). For seeds from several genotypes or seeds obtained by reciprocal crosses, grown without assimilate limitation, the mean volume of the cell cotyledons is highly positively correlated with the endoreduplication level in the mature cotyledon (Lemontey *et al.*, 2000) (fig. 1.42). Thus the control of the maximal cell size could be linked partially to the endoreduplication cycles in the cotyledonary cell.

Environmental effects

For a given seed cell number, the higher the temperature of the pod during the seed filling period is, the smaller is the mean maximal cell volume. Consequently the mean maximal cell volume depends on the environmental conditions during seed filling with a genotypic variability. The duration of seed filling expressed in degree-days decreased concomitantly with the elevation of temperature. Consequently the maximal duration of seed filling could not be considered as a genotypic constant.

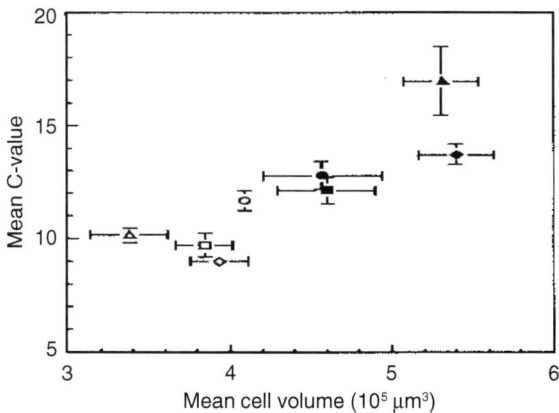

Figure 1.42. Relationship between the mean quantity of DNA and the mean cotyledon cell volume for different varieties and their reciprocal hybrids (from Lemontey *et al.*, 2000).

• Carbon and nitrogen resources exhaustion

Although the intrinsic limitation should be responsible for the termination of seed filling in some field conditions, the reserve exhaustion represents the most usual cause of termination of seed filling (Carrouée and Le Souder, 1992).

Synthesis of reserve compounds in pea seeds (starch ≈ Y 51 %; proteins ≈ 24 %; Carrouée and Gatel, 1995) depends not only on the instantaneous supply of carbon and nitrogen, but also on remobilization of nitrogen previously accumulated in vegetative parts. During the seed filling period carbon supply depends mostly on current photosynthesis (Warembourg *et al.*, 1984). Conversely nitrogen newly acquired during the seed filling period is unlikely to fulfill the high nitrogen demand of filling seeds, because both soil nitrate assimilation and nitrogen fixation decline during this period (Salon *et al.*, 2001): thus, nitrogen accumulation in seeds becomes metabolically closely associated with nitrogen remobilization. Nitrogen remobilization affects several nitrogen pools in vegetative parts (Peoples and Dalling, 1988). Rubisco which represents 50 % of the foliar proteins and is necessary for photosynthesis, constitutes a significant nitrogen source for remobilization (Matile 1992). Proteolysis and the export of nitrogen compounds lead to a decrease in the specific leaf nitrogen which affects the leaf photosynthetic activity (Sinclair and Horie, 1989). The reduction of photosynthetic capacity due to remobilisation and to a less extent the reduction of the leaf area at the end of the growth cycle (cf. chapter on C acquisition at the crop level) bring about the end of the seed filling period, when the carbon supply can no longer fulfill the seed's demand. The processes of remobilisation and their relationship with seed nitrogen demand are presented in the next chapter on seed protein concentration.

Summary

Seed weight determination is a complex phenomenon which depends on many variables:

- Cotyledon cell number. The cotyledon cell number determines the potential seed weight, because of its relationship with seed growth rate during seed filling and because it affects the potential maximal seed weight. Cell division occurs between flowering and the beginning of seed filling, so any stress occurring before the beginning of seed filling can affect the individual seed weight.
- Carbon and nitrogen supply available for filling seeds. Because seed growth rate determined by its cotyledon cell number is hardly affected by photoassimilate availability during the filling period, the reduction of photosynthetic activity caused by nitrogen remobilisation can end the filling period, and thus can limit the individual seed weight. Any biotic or abiotic stress during seed filling that depletes photosynthetic activity should lead to a reduction in the seed filling period.

Seed protein concentration

Annabelle Larmure, Nathalie Munier-Jolain

Seed protein concentration is one of the main quality criteria in pea. The mean value of the pea seed protein concentration (24% of dry matter) is intermediate between those of cereals (13% of dry matter) and soybean (40% of dry matter). For cultivated peas it varies greatly at harvest between years (fig. 1.43) as well as between batches sampled in a given year. This variability can be of genetic origin, but it is mostly due to environmental factors (Karjalainen and Kortet, 1987). Thus, for a given variety the variation in pea seed protein concentration is very large: for example for cv. Solara, from 17 to 28% of dry matter (Carrouée and Duchêne, 1993).

Seed protein concentration is calculated as the product of the seed nitrogen (N) concentration and a nitrogen to protein conversion factor determined from the amino acid composition of the seed proteins: 6.25 for pea (Carrouée and Duchêne, 1993).

In indeterminate plants like pea, the mean seed N concentration at harvest is the result of N and dry matter accumulation in seeds from different nodal positions on each stem. The N concentration of each seed is the ratio of:

— the amount of nitrogen in the individual seed—(the product of individual seed N accumulation rate and seed filling duration),

Protein concentration (% DM)

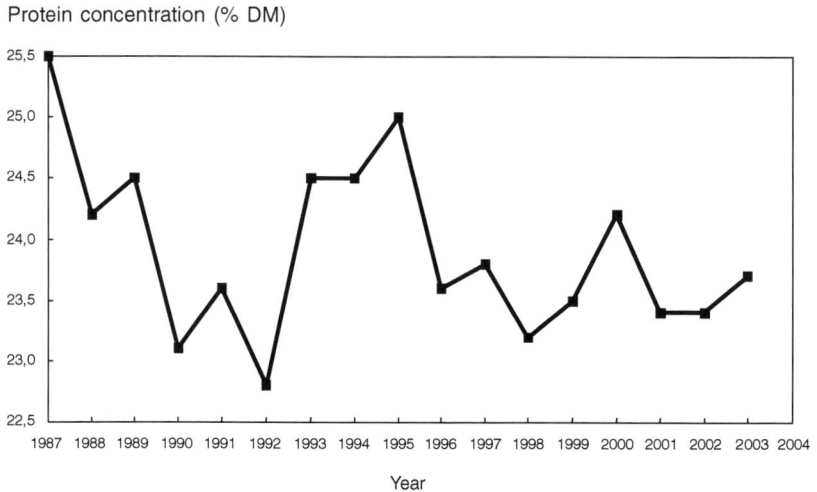

Figure 1.43. Variability against years of the mean seed protein concentration of pea in France. The number of batches analysed each year is about 180 and the standard mean error is 1,2 point each year (from UNIP data).

— the amount of dry matter in the individual seed—the product of individual dry matter accumulation rate (seed growth rate) and seed filling duration.

The duration of N accumulation and the duration of total dry matter accumulation seem to be equal for a given seed whatever the environmental conditions (Lhuillier-Soundélé *et al.*, 1999a). In order to understand the determination of seed N concentration it is necessary:

— to analyze how the seed growth rate and seed filling duration are determined (see p. 110 and 114–115)
— to analyze how the individual seed N accumulation rate is regulated at the different reproductive nodes. It is also essential to understand the determinism of seed N accumulation rate over the course of time in order to explain the variation in seed filling duration caused by nitrogen remobilization from vegetative parts to filling seeds (see previous chapter on individual seed weight).

Differential regulation of individual seed N accumulation rate and individual seed growth rate

During the seed filling period, seed growth rate at a given node is unaffected by variation in assimilate availability over time (Munier-Jolain *et al.*, 1998). In contrast, the individual seed N accumulation rate is affected by an increase or a decrease in N assimilate availability (table 1.4; Lhuillier-Soundélé *et al.*, 1999a).

Modifying the N source/sink ratio and the location of source and sink organs in the plant has also shown that there is no hierarchy in the allocation of N between seeds of different ages and node levels (Lhuillier-Soundélé *et al.*, 1999b). Thus, the seed N accumulation rate at the different nodes is unaffected by the uneven plant N distribution. These results indicate that seed growth rate and seed N accumulation rate are differently regulated:

— filling seeds have no priority among competing sinks over N assimilate allocation, whilst they do have priority over the allocation of photoassimilates,

— seed N accumulation rate varies during the seed filling period according to plant N availability, whereas seed growth rate is fixed at the beginning of seed filling. Seed N accumulation rate can be considered as identical for all seeds at a given moment whatever the nodal position or the growth rate of the seed.

Table 1.4. Effects of treatments which modify the nitrogen assimilate availablity during the seed filling period on the nitrogen accumulation rate in individuel seed (treatments which modify the nitrogen assimilate availablity were realised during the seed filling period of the seeds at the studied node).

Treatment	Effect on the nitrogen availability	Nitrogen accumulation rate in individual seed (µg/seed/degree-day)
Control	↘	20,5
Non-nodulant		11,7
Control	↗	27,2
Nitrogen supply		34,8
Control	↗	27,1
Pod ablation		37,9
Leaf ablation	↘	16,8

Characterization of the amount of N available to filling seeds

The N accumulated in seeds during the seed filling period has two origins: exogenous N being assimilated and/or fixed by the plant and remobilized N from N source organs (fig. 1.44).

• Exogenous N accumulated during the seed filling period

The N accumulated by the plant during the seed filling period significantly contributes to the total N accumulated in seeds at harvest; thus 25 to 40 % of the total amount of N in a plant can be accumulated during the seed

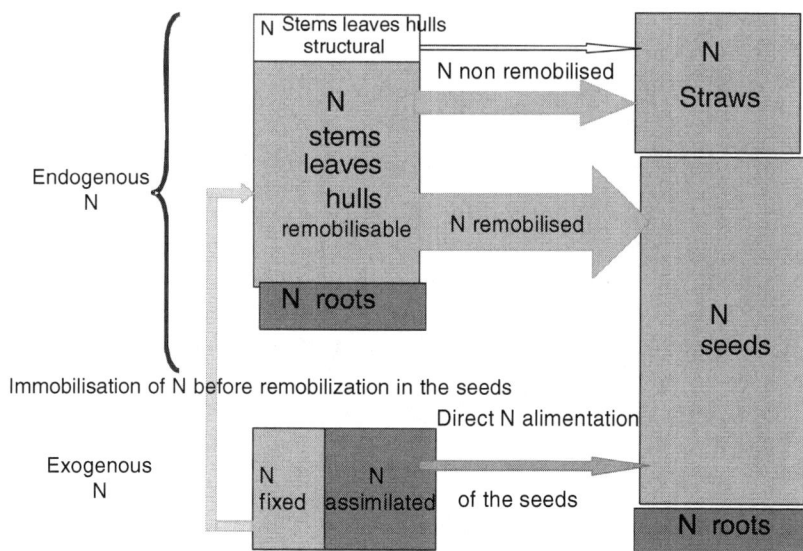

Figure 1.44. Nitrogen (N) flow in a pea plant during the seed filling period.

filling period (Lhuillier-Soundélé, 1999; Crozat *et al.*, 1994). Nitrogen nutrition during seed filling relies on both symbiotic fixation and soil mineral N absorption. Even though symbiotic fixation decreases during the seed filling period, it seems to be still effective and to stop only at physiological maturity (Sparrow *et al.*, 1995; Voisin *et al.*, 2002). Soil mineral N absorption can continue during seed filling; however it is limited by the availability of N in the soil.

The N accumulated by the plant during the seed filling period seems to be allocated to seeds first. In fact all the N accumulated by the plant during seed filling is accumulated in the seeds (Sagan, 1992). However, the physiological mechanisms controlling nitrogen fluxes in pea plants are complex.

The transfer to the seeds of the N accumulated by the plant seems to be a slow process (Pate and Flinn, 1973). Thus, the N accumulated by the plant seems not to be immediately available, but allows the release of an equivalent amount of N to the seeds by protein degradation,

- Remobilized N

Source organs and structural nitrogen: leaves, stems and pods are the major contributors to the N supply of seeds. Roots were shown to provide only about 3 % of the N remobilized to the seeds (Peoples and Dalling, 1988).

However, a proportion of N in the source organs is not available for remobilization to the seeds because it is associated with structural components (Caloin and Yu, 1984). The amount of non-remobilizable N depends on the concentration of non-remobilizable or structural N (Munier-Jolain et al., 1996). In pea, this structural N concentration was estimated at 7.2 mg.g^{-1} of dry matter; however this value could vary with the organ and the N nutrition level of the plant (Lhuillier-Soundélé et al., 1999b; Larmure and Munier-Jolain, 2004).

Remobilization process: remobilized N from N accumulated in N source organs before the beginning of seed filling contributes 60 % to the final N yield. However, ^{15}N labeling of the exogenous N accumulation pathway has shown that 90 % of the N accumulated in the seeds comes from remobilization (Grandgirard, 2002): remobilization processes seem to involve both N accumulated before the beginning of seed filling and exogenous N accumulated during the seed filling period.

The rate of available N remobilization to the filling seeds seems to be regulated in different ways over the course of time: it could be limited at first by exogenous plant N accumulation and the setting up of protein degradation mechanisms, while later, at the end of the seed filling period, it could depend purely on the amount of N available per seed (Grandgirard, 2002, Salon et al., 2001).

- Influence of the temperature on the amount of N available to seeds

Temperature during the seed filling period may influence plant N partitioning (Spiertz, 1977). Thus, seed N concentration decreases when air temperatures decrease during the seed filling period (Larmure et al., 2005). This result could be linked to a reduction in N availability to filling seeds. The few results available for pea or soybean suggest that neither exogenous N accumulation nor the rate of N remobilisation are affected by temperature variation over a moderate range of 13–29°C (Grandgirard et al., 2001; Larmure et al., 2005). However, competition between seeds and vegetative sinks could be modified at the lower end of this moderate range and with constant radiation as a consequence of an increase in plant carbon accumulation expressed in degree-days. The carbon flux allocated to seeds is not modified because of the priority of carbon partitioning to filling seeds, so the resulting additional C flux allows new competing vegetative sinks to grow. The increase in vegetative growth seems to attract part of available N at the expense of filling seeds (Larmure et al., 2005).

Determinism of the individual seed N accumulation rate

- The amount of N potentially available to filling seeds at a given time can be estimated using the amount of exogenous N accumulated by the plant and the amount of remobilisable N. The characterization of the amount of N available to filling seeds provided a relationship between the rate of individual seed N accumulation and plant N availability (amount of N available per filling seed) (fig. 1.45;

Speed of nitrogen accumulation of a seed$_{(t\ to\ t+1)}$ (μg N. seed^{-1}; °Cd^{-1})

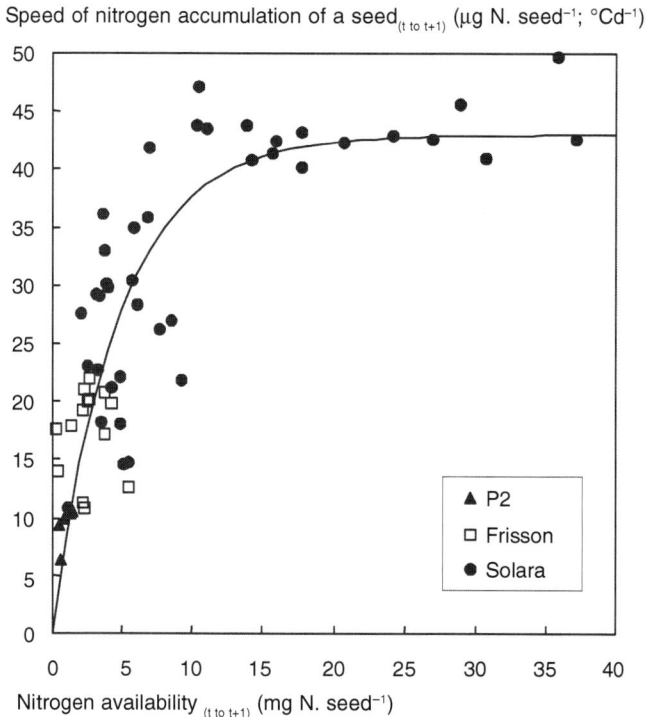

Nitrogen availability $_{(t\ to\ t+1)}$ (mg N. seed^{-1})

Figure 1.45. Relationship between the speed of nitrogen accumulation of a seed and the nitrogen availability in the plant (from Lhuillier-Soundélé *et al.*, 1999b, reprint with the authorization of the American Society of Agronomy, ©ASA, Crop Sci., 39, 1741-1748).

Lhuillier-Soundélé *et al.*, 1999b). This relationship is unique for the three genotypes studied (P2, Frisson, Solara). It demonstrates that the individual seed N accumulation rate increases with an increase in plant N availability until reaching a maximum rate of individual seed N accumulation.
- Partial depodding experiments leading to large increases in N availability to the remaining filling seeds allowed the estimation of

the value of the maximum rate of individual seed N accumulation for three genotypes (Frisson, Solara and Baccara). This study demonstrated that the maximum rate of individual seed N accumulation varies among genotypes in accordance with the individual seed weight (Larmure and Munier-Jolain, 2004).

Conclusion

- Seed protein concentration of a pea crop calculated from the seed nitrogen (N) concentration (x the conversion factor 6.25) is the ratio of the seed N accumulation rate and the seed growth rate. A high seed protein concentration can be the result of a high mean seed N accumulation rate and/or a low seed growth rate. These two rates are mainly determined by independent mechanisms. The individual seed N accumulation rate is similar at a given time for all filling seeds of the crop and depends on plant N availability. It is likely to decrease in the course of the seed filling period with the decrease of the amount of N available to filling seeds (exogenous N and remobilisable N). However the seed growth rate is fixed at the beginning of seed filling and depends on cotyledon cell number. A genetic or environmental relationship probably exists between the seed N accumulation rate and the cotyledon cell number, as both variables depend on both genotype and N nutrition level.
- The knowledge about the seed N accumulation rate has been synthesized in a model simulating N partitioning to seeds during the seed filling period in pea (Larmure and Munier-Jolain, 2004). This model, which was intended to be introduced into a global model of pea crop growth, will be an important component to simulate seed protein concentration, the N concentration of vegetative parts and the date of termination of seed filling in pea.

Elaboration of germination quality: the determinism of seed fragility

Jean-Albert Fougereux, Thierry Doré, Sylvie Ducournau, Fabienne Ladonne

The germination quality of seeds is particularly important in large-seeded leguminous plants, especially peas, for two essential reasons:

— the high cost of the seeds: although the sowing density for the pea crop is on average four times lower than that of wheat for example, its thousand-grain weight is six times higher; the weight of seed

required to sow a pea crop is thus 1.5 times greater than for wheat, and the cost of seeds is 1.8 times greater because of the higher cost of the "raw material".
— the difficulty in controlling germination quality. Pea seeds are subject to certain problems which are not always easy to control in seed production.

The components of germination quality

The germination quality of a seed lot can be defined as all those properties of the lot which make it possible to obtain after sowing a healthy and uniform crop over a broad range of soil and weather conditions. In peas, this germination quality in a broad sense is the resultant of three principal components: physical, pathological and physiological components.

• Physical quality

Pea seeds are notoriously brittle. The damage caused by shocks at harvest or handling results in broken seeds ("visible breakage"), which can be eliminated by sorting, but also in reduced germination due to the appearance of abnormal seedlings with damaged embryos ("invisible breakage"), impossible to correct by sorting. The physical quality of the seeds is regarded as the principal factor affecting the germination quality of seed lots.

• Pathological quality

The pea seed can be host to many cryptogams. In general, inoculation occurs on the mother plant when weather conditions are favourable. After harvest, the parasite is found, like the seed, in the quiescent state. It is at the time of imbibition that it starts again to be propagated. According to the nature of the parasite and the level of infestation, the damage is variable. It often results in necrotic spots or rots, accompanied by a deterioration of the growth of the seedlings, and can in extreme cases result in the death of the seed. In France, a dataset relating to more than 2500 seed batches produced between 1987 and 2003 shows that the diseases most frequently met are, in descending order of frequency: *Botrytis* spp. (73 % of contaminated batches), *Fusarium* spp. (71 %), *Mycosphaerella pinodes* (49 %), *Stemphylium botryosum* (35 %), *Penicillium* spp. (21 %), *Phoma medicaginis* var. *pinodella* (16 %), *Ascochyta pisi* (12 %). Mildew (*Peronospora pisi*), not studied in this dataset, is also commonly present.

These cryptogamic diseases can all affect seed germination to different extents. Contamination is generally only superficial, and usually seed dressing with fungicidal treatment limits the pathogenic capacity of the fungi and maintains a satisfactory germination capacity. However in some years, in spite of fungicidal protection of the mother plant and appropriate seed dressing, a considerable percentage of seed lots must be eliminated due to inadequate germination.

Germination quality of pea seeds can also be affected by two insects: the pea beetle *Bruchus pisorum* and the pea moth *Cydia nigricana*. A pea seed parasitized by the beetle has only one chance in three to produce a normal seedling; one in two in the case of the moth. Sorting eliminates only a fraction of the parasitized seeds, and often with significant waste. In the areas concerned, the maintenance of the germination quality of seed lots requires specific insecticidal programs during growth.

- ## Physiological quality

Defects of physiological origin are more difficult to define. One can regard this category as all the germination defects which are neither of pathological nor mechanical origin. They correspond to a dysfunction of the seed leading to deformations or imbalances of growth of the seedlings. As for the defects of pathological origin, they can involve, in the most extreme cases, the death of the seeds. The state of the cellular membranes of the embryo (embryonic axis and cotyledons) constitutes one of the first factors responsible for these physiological defects (Powell, 1985).

Elaboration of germination quality: determinism of seed fragility

Only the elaboration of physical quality of the seeds is treated here. It depends mainly: (i) on factors related to the harvest conditions, and (ii) of the ability of the seed lot to resist mechanical injury at harvest, usually called "fragility", which is very variable from one crop to another, or even within a single crop. The search for the optimum conditions at harvest was the object of experiments undertaken by the FNAMS (for example, THIBAUD, 2004; GNIS & FNAMS, 1992). We will be interested here mainly in the study of the causes of fragility of the seed lot before harvest. The majority of work quoted hereafter received the support of the French Ministry of Agriculture. The program was carried out in partnership with INA-PG, the SNES, ARVALIS-Institut du Végétal—and seed companies. This work also lies within the scope of the specific actions of the GNIS.

- ## Measurement and definition of seed fragility

Although important, harvest conditions alone cannot explain all the forms of mechanical damage observed: under similar harvest conditions, there is much variability in the brittleness and behaviour of the seed lots. In order to quantify this fragility before harvest, the FNAMS developed an instrument, the fragimeter, to measure the ability of seed lots to resist shocks (Fougereux, 1999a): the fragility of a seed lot is thus defined by the percentage of broken seed (by weight), compared to the initial weight of the sample. High values of fragility correspond to seed lots producing many broken seeds after passage in the fragimeter. It is then said that the seed lot is very fragile.

• Water concentration at harvest: A strong incidence on Fragility

Seed water concentration is the first explanatory factor of the variations noted between seed lots for the sensitivity to mechanical damage (MERIAUX, 2002). Indeed, the loss of germination at harvest is strongly related to seed water concentration (fig. 1.46). In the laboratory, it is shown that the lower the water content, the more fragile are the seeds. This effect of the water concentration is more or less marked according to seed lots.

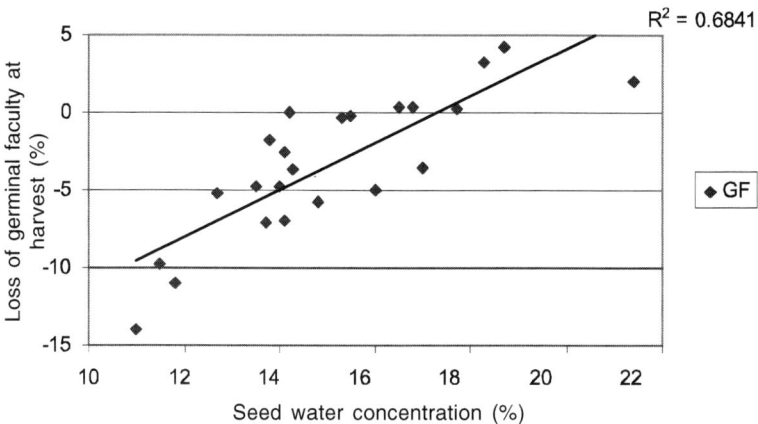

Figure 1.46. Relationship between seed water concentration and loss of germinal faculty (GF) at harvest. West Center/Parisian Bassin Areas—batches of production 1989.

• Other internal factors affecting seed fragility

The water concentration is not the only explanatory factor of seed fragility. While working at constant seed water concentration, one still notes much variability in fragility, which could be explained by various parameters.

Thus, in an investigation carried out from 1995 to 1997 in two areas of France, the variability of seed fragility before harvest was characterized on manually collected seed lots, while working with constant moisture of 12 % (Fougereux et al., 1999a; Fougereux, 1999b). For the two varieties studied, results show very great variability (fig. 1.47).

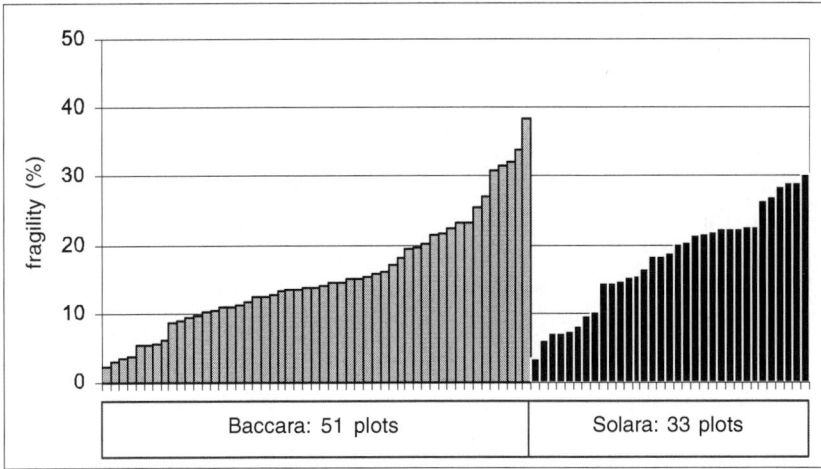

Figure 1.47. Variability of fragility observed at harvest for 84 pea plots. Seeds balance to equilibrate to 12 % humidity—Survey Beauce/Aunis 1995 to 1997.

• Varietal sensitivity to fragility

The variability of intra-varietal fragility appears much greater than that from one variety to another. Hence it did not appear important until now to seek to establish a true varietal classification for this factor. Moreover, a classification of the varieties would require considerable resources for characterizing many seed lots for each variety. Nevertheless a varietal effect was noted for certain old varieties (FNAMS, 1997). There is no classification for this factor for the varieties currently cultivated in France.

• Fragility and physical characteristics of seeds

Apart from moisture, it was observed that certain physical characteristics of seeds seem dependant, partly at least, on their fragility (Fougereux *et al.*, 1998b).

Seed weight
For a large number of seed lots with constant water concentration, only a weak positive correlation appears between thousand grain weight (TGW) and fragility. The low TGW seed lots have a very variable fragility. On the other hand, the high TSW seed lots never exhibit very low fragility.

The percentage of tegument
The lower the percentage of tegument in the seed (dry weight of seed tegument x 100 /total dry weight of the seed), the more fragile is the seed. However, it is difficult to assess the relative importance of seed weight and

percentage of tegument, because high TGW is associated with small percentages of tegument.

Tegument/embryo adherence—adherence between the cotyledons

The observation of pea seeds at harvest shows, for a given variety, that the seeds can have a very variable appearance. Even on the surface of a single seed, very marked nuances of colouring sometimes occur. After seed tegument removal, these nuances of colouring appear to depend on whether or not the tegument adheres to the embryo. The zones where the tegument adheres to the cotyledons are definitely darker than the zones where there is a gap between these two compartments. In addition, when teguments and cotyledons of a seed adhere on only a small area of the seed, the two cotyledons are generally fused together.

However adherence between cotyledons is rather strongly related to fragility (fig. 1.48). By adding this effect with the average weight of the seeds, 66 % of the variability in fragility is explained (at constant moisture concentration). Figure 1.49 summarizes the current assumptions concerning the relations between the fragility of a seed lot and adherence between the various compartments of the seeds.

- ## Study of the effects of soil and weather on fragility from a crop survey

Effect of maturity state at harvest

One of the main factors likely to explain the variations in fragility is the level of maturity of the seeds at the moment of sampling. Regardless of the water concentration at the time of the shock, the seeds gradually become more brittle towards the end of growth. This process can occur at varying rates.

Other soil and weather factors

For seed lots which have not reached a very advanced stage at harvest (water concentration > 15 %), fragility tends to increase with the temperature during the period [beginning of flowering—end of limiting stage of seed abortion], which corresponds to the period of embryogenesis (fig. 1.50). This was the only soil/weather factor to emerge from the investigation.

- ## Effect of rain at the end of plant growth and the effect of drying conditions

Simulated rainfall at the end of plant growth

Rainfall simulation was performed in the field at the end of the plant cycle, with several stages of sprinkling (at seed moisture contents of 20 %,

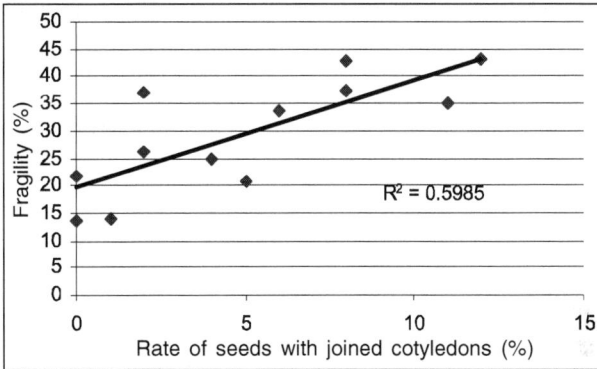

Figure 1.48. Relationship between seed fragility and the rate of seeds with joined cotyledons. (8 batches of cv. Baccara, 100 seeds observed by batch).

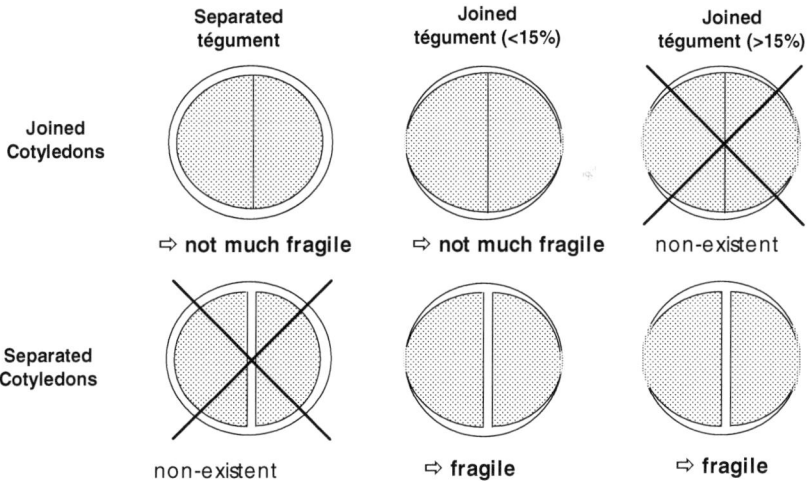

Figure 1.49. Summary of relationships observed between seed fragility and adhesion of its différent compartments.

15 %, or no sprinkling). These experiments showed that rain at the end of seed cycle can have a significant effect on the brittleness of the seeds (data not shown).

Moistening experiments in the laboratory

A laboratory experiment showed clearly that the seeds became more brittle when they had been moistened beforehand and then dried back to the initial moisture content (table 1.5).

The influence of various drying conditions after re-hydration of the seeds to 25 % water concentration was also tested: the hottest and fastest drying conditions lead to particularly fragile seeds.

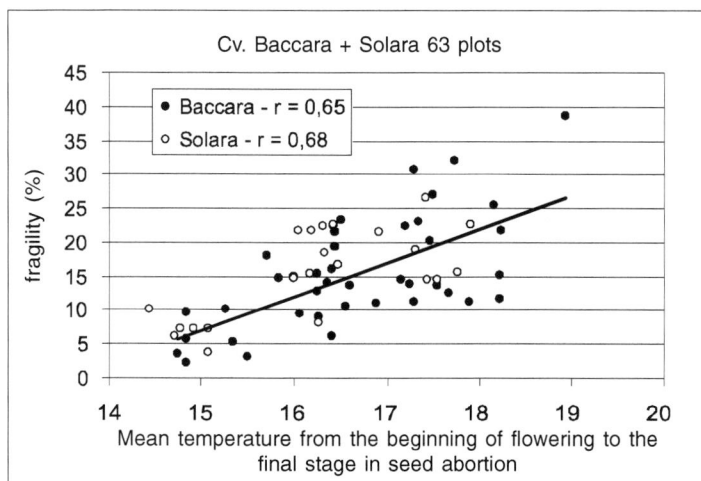

Figure 1.50. Relationship between fragility and mean temperature from the beginning of flowering to the final stage in seed abortion (plots with extreme water concentration at harvest have been eliminated).

The application of various drying conditions to seeds taken directly from the plant at maturity makes it possible to generate different durations and speeds of drying (table 1.6).

The drying conditions strongly influence fragility, with a rate of breakage which varies between 10 and 60 % after a fragility test:

Table 1.5. Study of the causes of seed fragility in pea—Laboratory experiment of moistering seeds at 12% water concentration. Data: FNAMS 1995.

Treatment	Water concentration after moistering %	Fragility % (after stabilisation at 12% water concentration)
M1 (control)	12.3	19
M2	16.7	59
M3	20.5	53
M4	25.2	48

Table 1.6. Effects of different drying conditions on seed fragility in pea. Data: FNAMS-1996, cv. Baccara—Seeds have been taken at node 1 of the plant when it was at 57% water concentration.

Drying conditions	Duration for loss of 50% water concentration (hour)	Fragility (%)
15°C-75% WC	40	10.25
25°C-75% WC	50	15.75
35°C-75% WC	32	22.25
25°C-WC<20%	23	60.25
25°C-WC>95%	82	12.5

— high drying rates lead to more fragile seeds;
— at a given drying rate, higher temperatures seem to lead to more fragile seeds;
— the effect of the drying conditions depends on the maturity of the seed at sampling.

Conclusion

Although knowledge relating to seed fragility in pea and its determinism has advanced significantly during the last ten years, possibilities for the control of this fragility remain limited. Neither the temperature during embryogenesis nor the drying conditions are under the farmer's control. It is primarily by the control of the stage of harvest and by the implementation of optimal harvesting techniques that one can hope to preserve the physical integrity of the seed, as is necessary for obtaining seed lots of good germination quality.

PART II

Analysis of the effects of abiotic and biotic stresses

Abiotic stresses

Influence of water deficit on pea canopy functioning

Jérémie Lecoeur and Lydie Guilioni

Field survey networks or multifactorial analyses of yield variability have shown (Pumphrey *et al.*, 1979) that water deficit is without question a major limiting factor in pea. The aim of this chapter is to synthesize the main results relating to the analysis of its effects on pea crop physiology. The first part defines the terms of deficit, constraint and water stresses relating either to the concept of water availability in the environment or to plant responses. Changes in different water status variables in response to soil drying are described for the most frequent water scenarios in the main pea cropping areas. The nature of plant responses is discussed in the second part, where the effects of water deficits are reported on growth, development, and yield and quality elaboration.

Characterization of a water deficit and its perception by the plant

- Notions of water deficit, water constraint and water stress

There is a certain amount of confusion between these above three terms although they define distinct notions.

A water deficit occurs when the water available for the plant does not allow it to respond to the climatic demand. In this case, canopy evapotranspiration is below maximal evapotranspiration. The term water deficit appears to be relatively generic in relation to the question of the plant water supply as it refers to the balance between water availability in the environment and plant needs. These needs result from the combination of both climatic demands and plant structure. The notion of water deficit is therefore a physical concept which does not include the physiology of the plant. In terms of water supply, the plant is primarily a means of transferring water between the soil and the atmosphere since the proportion of water stored in the plant compared to the proportion which is transpired is in the order of one percent. A water deficit can therefore be a result of low water availability in the soil portion explored by the roots and/or a heavy evaporative demand at leaf level. But as pea is generally cultivated in temperate zones where climatic demand seldom exceeds 6 to 7 mm per day, the analysis of water deficit effects refers rather to the edaphic water deficit, i.e. it is linked to a reduction in soil water availability.

The terms water constraint and water stress refer to two different notions related to the nature of the plant response with regard to the intensity of the water deficit. There is water constraint when plant functioning is affected by the water deficit without a reduction in tissue water content. Finally, there is water stress when plant tissues experience a fall in water content which affects the whole plant metabolism. Water constraint thus always precedes water stress.

• A gradual plant response to soil drying

Presently, there is no universal variable to characterize the plant water status. This characterization can be approached from the angle of the water status of the plant itself or the water fluxes within it, which is generally assessed by means of its transpiration level. The plant water status can be characterized either by measurements based on the energy state of water in its tissues like the foliar water potential or on the water content of its tissues like relative water content. Plant transpiration level is most commonly assessed by measuring stomatic conductance. The majority of the variables used to characterize the perception of the water deficit by the plant are quantitatively linked with the soil water status (Lecoeur, 1994a; Combaud, 1996). Soil water status thus appears to be a good indicator of the water deficit experienced by the plant. The effect of the water deficit will be expressed below through relationships between the variables describing the response of the plant and the FTSW[1]. This variable has the advantage of being easily estimated by a water balance (Sarr et al., 2004)[1].

---

1. Fraction of transpirable soil water or FTSW: This variable corresponds, for a given soil volume explored by roots, to the ratio between the actual amount of available soil water at a given time and the maximal amount of available soil water. The maximal amount of available soil water corresponds to the difference of soil water between the field capacity and the soil water content when the stomatal conductance of the plant reaches zero or when the plant transpiration is less than 10 % of the transpiration of a plant which does not experience a water deficit.

As soil gradually dries, the combined use of the different variables makes it possible to identify a succession of events which characterize the response of a pea plant to a water deficit.

Plant water status in absence of water deficit

Before plant responses to soil drying can be described in terms of water status, it is necessary to characterize the average water status of a plant in the absence of water deficit. The average stati described here are the results of field experiment trials monitoring in the Montpellier region (Lecoeur, 1994a; Combaud, 1996). They correspond to situations where the fraction of transpirable soil water (FTSW) is greater than 0.6 (Lecoeur and Sinclair, 1996):

— the basic potential (Ψ_{base}) of a pea plant when soil is close to saturation ranges between –0.1 and –0.2 MPa. The leaf water potential at solar midday ranges between –0.3 and –0.7 MPa between emergence and the beginning of flowering, depending on environmental conditions. It then falls gradually to –1.0 MPa during reproductive development, probably in connection with leaf ageing;

— the leaf water concentration is stable during the cycle with an average value of 85 % (90 % of values range between 82 and 86 %). Water concentration at water saturation is 87 %. This value can be regarded as a characteristic of the species, linked to foliar morphology.

— the stomatal conductance at solar midday ranges between 0.3 and 0.5 $m^{-2}.s^{-1}$. Abscisic acid content (ABA) in the xylem sap, a hormone which signals water deficit, ranges between 6 and 10 $ng.ml^{-1}$ (Lecoeur, 1994a).

— in the absence of edaphic water deficit and in response to variable climatic demands, stomatal conductance presents greater variability than tissue water content. This indicates an isohydric plant behavior where the regulation of fluxes supports the preservation of the tissue water status to the detriment of gaseous exchanges.

Plant responses to a progressive soil drying

As the soil dries gradually, it is possible to identify periods in plant perception and response to the water deficit by means of changing water variables (Lecoeur, 1994b). This typology has, however, no cognitive value, it merely makes it possible to identify bench marks in the gradual plant response to soil drying:

— the first stage could be described as *early water deficit perception*. It occurs when the fraction of transpirable soil water (FTSW) falls below 0.4 (Lecoeur and Sinclair, 1996). It is characterized by an increase in ABA content of xylem sap which exceeds 15 ng.ml⁻¹ and a reduction in the stomatic conductance (fig. 2.1a) and in the expansion rate of the vegetative organs (fig. 2.1b) proportional to the level of soil drying;

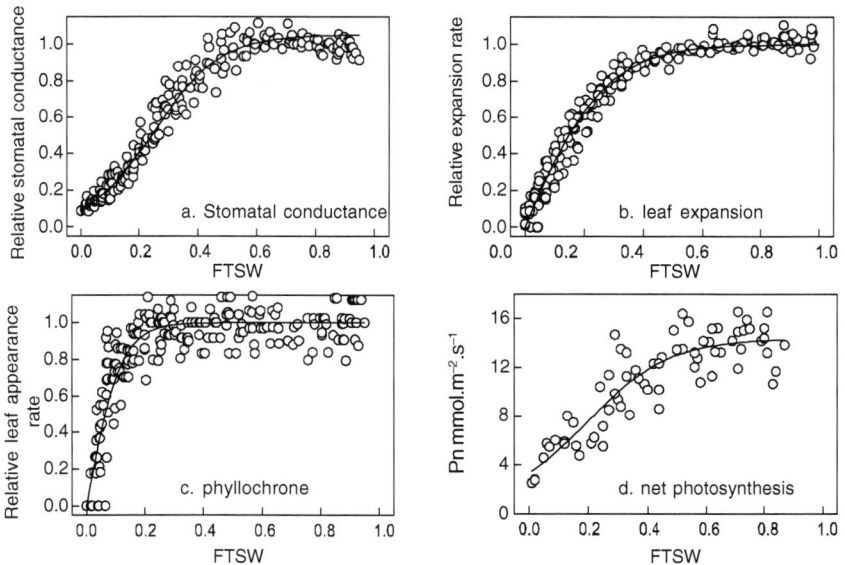

Figure 2.1. Change in relation to fraction of transpirable soil water (FTSW) of (a) stomatal conductance. Expressed in relation to the FTSW (see Lecoeur and Sinclair, 1996a). (b) leaf expansion rate. Expressed in relation to leaf expansion rate of well-watered plants (see Lecoeur and Sinclair, 1996a). (c) leaf production rate. Expressed in relation to leaf production rate of well-watered plants (see Lecoeur and Guilioni, 1998). (d) net photosynthesis (see Guilioni *et al.*, 2003).

— the second stage corresponds to the notion of *water constraint*. Stomatic conductance halves compared to the case of a well-watered plant with values ranging between 0.15 and 0.25 mol. m⁻².s⁻¹. Foliar water potentials at solar midday start to fall with values of around –1.0 MPa. On the other hand, there is no drop in turgescence or in relative water concentration because of an effective osmotic adjustment. FTSW ranges between 0.4 and 0.2.

During these two first stages, only the plant water fluxes are affected. The plant water status remains the same as a well-watered plant.

— the third stage or *moderate water stress* marks the beginning of the modification of the physiological status of plant tissues. Stomatal

conductance falls further to reach values of around 0.1 mol.m^{-2}.s^{-1} and leaf water potentials at solar midday reach –1.2 to –1.4 MPa. The ABA content of xylem sap exceeds 30 ng.ml^{-1} and an accumulation of senescence-specific ARN can be observed (Pic et al., 2002). Osmotic adjustment is no longer sufficient to maintain turgescence and the relative tissue water concentration falls to values of approximately 75 %. Tissue expansion is very low or ceases. FTSW ranges between 0.2 and 0.1.
— the fourth stage marks the beginning of tissue dehydration tolerance processes, corresponding to a severe water stress. The stomates are completely closed. The plant still loses a little water by cuticular transpiration which is approximately 0.05 mol.m^{-2} s^{-1}. The leaf water potential at solar midday reaches minimal observed values below –1,6 MPa. There is no more turgescence and relative water concentration is below 70 %. Transpirable soil water has been exhausted. In this situation, depending on climatic demand, the plant could die in a few hours from the dehydration of these tissues.

• Occurrence of water deficits in France

Pea crops in France are never in a situation of optimal water supply with the exception of specific experimental situations. In any case, this situation is undesirable for production because it involves excessive vegetative development (Turc et al.,1990). According to the typology defined above (cf. 1.2), a crop experiences at the very least mild and temporary phases of water deficit. The most frequent events are water constraints and moderate water stresses during reproductive development. Earlier events and severe water stresses are uncommon. Besides, the latter are not compatible with financially viable production because severe water stresses can heavily reduce yield. The most commonly observed sequence of events is an absence of water deficit until flowering, water constraints or moderate water stress between flowering and the end of the abortion limit stage with a return to the optimum, then the gradual implementation of severe water stress during seed filling until plant maturity.

Effect on development

• Floral initiation and beginning of flowering

Water deficit does not appear to affect floral initiation and the number of phytomers at flowering. An experiment conducted in greenhouses with various levels of soil drying maintained constant between emergence and physiological maturity of the plant did not result in a modification of the date of floral initiation, nor of the number of phytomers at flowering (Guilioni and Lecoeur, Np). Moreover, the probability of water deficits occurring in the field between emergence and flowering is low in most areas of production. Floral initiation can therefore be considered as being independent of the plant water supply conditions.

• Progression rate of phenological stages

The progression rate of phenological stages along the stems is remarkably stable in response to water deficit. For vegetative development, the rates of phytomer initiation by the meristem (plastochrone) and of leaf appearance (phyllochrone) are not affected by a moderate water deficit for which the FTSW is greater than 0.2 (Lecoeur, 1994a; Combaud, 1996). On the other hand, a severe water deficit reduces the rate of phytomer initiation and appearance of leaves. This reduction is proportional to the level of soil drying for FTSW ranging between 0.2 and 0 (fig. 2.1c) (Lecoeur and Guilioni, 1998). This means that apart from extreme situations of water deficit, the rate of phytomer initiation and appearance of leaves depend only on air temperature.

Similar results were obtained for reproductive development. Progression rates of flowering, final stage in seed abortion and physiological maturity are not affected by a moderate water deficit (Ney et al., 1994; Lecoeur, 1994a). Severe deficits can slow down the progression rate of reproductive development stages (Lecoeur, 1994a). Flowering progression rate appeared to be more sensitive than other development stages. However, this difference is only expressed for very intense water deficits. Development rates are thus only reduced for significant levels of soil drying.

In addition, the production of ramifications is seldom affected by water deficit (Combaud, 1996) because their number is determined before the appearance of most water deficits. The organ production process at canopy level appears to be relatively insensitive to water deficit. This explains why pea development is regarded as being insensitive to water deficit in farming practice.

• Production duration of vegetative and reproductive organs

A water deficit brings about an anticipated stop in the production of new phytomers by the cauline meristem, thus reducing the final number of reproductive phytomers and shortening the cycle. The reduction in the final number of reproductive phytomers is all the greater, the earlier and more intense the water deficit appears (Lecoeur, 1994a; Guilioni et al.,

2003). Thus, for a reference situation with a final number of reproductive phytomers of 10, a moderate water deficit before or after flowering would on average halve the number of reproductive phytomers, a severe deficit before flowering would reduce the number by 3 to 4 and a severe post-flowering deficit by 4 to 5.

Little is known about the cause of this cessation in producing phytomers, including in non-limiting water conditions. It could be a result of trophic competition between the apical tips and developing seeds (Jeuffroy and Warembourg, 1991). In this hypothesis, the trophic priority of seeds causes the meristem to stop when quantities of assimilates are insufficient to satisfy all the needs. The water deficit causes an earlier stop of the apical meristem functioning by modifying the sources-sinks relationships. It reduces vegetative organ size (see fig. 2.1) and number of reproductive organs. It also affects their physiological state. This in turn leads to an increase in demand for assimilates of the first reproductive phytomers in a context of diminished availability. This reduction in assimilates availability is much greater for post-flowering water deficits which result in reductions in both leaf area and photosynthetic activity. The increased pressure of the first reproductive phytomers on the pool of available assimilate leads to an earlier limitation of assimilates supply to new phytomers and the meristem. This earlier limitation leads to a shortening of the production duration of phytomers and to a reduction in the final number of reproductive phytomers.

Effect on growth

- Expansion of the vegetative organs

A water deficit, whatever its intensity, reduces the size of all developing vegetative organs on the plant at the time of its occurrence, including on the microscopic organs in their early stages in the apical bud (Lecoeur *et al.*, 1995). The effects of a specific water deficit on final organ size will thus be manifested for several weeks after the deficit end, until the time when all the affected organs have completed their expansion.

A water deficit affects the expansion of an organ through a reduction in cell wall plasticity and the reduction in pressure of intracellular turgescence (Cosgrove, 1993). The first effect is linked to changes in mechanical properties of the walls, partly under the control of a root hormonal signal like the ABA content of xylem sap (Davies *et al.*, 1994). The second is a hydraulic effect linked to the reduction in cell water content (Lockhart, 1965). These effects lead to a reduction in expansion rate of tissues through a reduction in rates of cellular expansion and division. In the case of moderate water deficits, the reduction in cell division rate is a consequence of the reduction in expansion rate. It is not a question of a direct effect on the cellular cycle. For severe water deficits, this reduction is a result of a general deterioration in cellular metabolism. The effect of the water deficit on the expansion rate of organs can be represented by a simple relationship between the reduction in expansion rate relative to a control without water deficit and the soil water status characterized by the FTSW (fig. 2.1b) (Lecoeur and Sinclair, 1996a). This relationship indicates that as long as the FTSW is greater than 0.4, organ expansion rate is maximal. It then decreases with the soil water status to cancel itself out when the FTSW is equal to 0.

A mechanistic model on a phytomer scale was proposed to predict the effect of a water deficit on leaf area (Lecoeur and Sinclair, 1996b). This model estimates the reduction in leaf area relative to well-watered plants as the product of the number of epidermic cells by the average area of these cells. The development of each leaf is broken down into two phases; a first phase of cell division corresponding to the first three-quarters of foliar development and a second phase of cell expansion corresponding to the final third of foliar development. These two periods overlap for a short time (between 66 % and 75 % of leaf development) which appears to be the leaf expansion phase most sensitive to water deficit. The same relationship links reductions in cellular division and expansion rates to the FTSW (Lecoeur and Sinclair, 196a). This model can predict dynamic changes in the area of each leaf over time and final leaf area for a broad range of pre- and post-flowering water deficits with moderate or strong water constraints. It can predict complete profiles of leaf area in the field for main stems and ramifications (Combaud, 1996).

In the field, reductions in organ size can reach 80 % of the size of well-watered plants. In this case, the water deficit has to be intense and prolonged in order to affect both the number and the area of epidermic cells (Lecoeur et al., 1995). A water constraint or a moderate water stress leads to an average reduction in leaf area of between 20 % and 40 %. A water deficit, which affects all expanding organs at the time it occurs, affects approximately ten phytomers (fig. 2.2a). Thus, it changes completely the distribution of the leaf area or internode length along the stem.

• Photosynthesis and biomass production

Biomass accumulation is primarily a result of photosynthetic activity. According to the intensity of the water deficit, different processes can be affected. The first is CO_2 supply to the carboxylation sites in the chloroplasts. It can be limited in the case of the early perception of the water deficit. It results from partial closure of stomates in response to hormonal

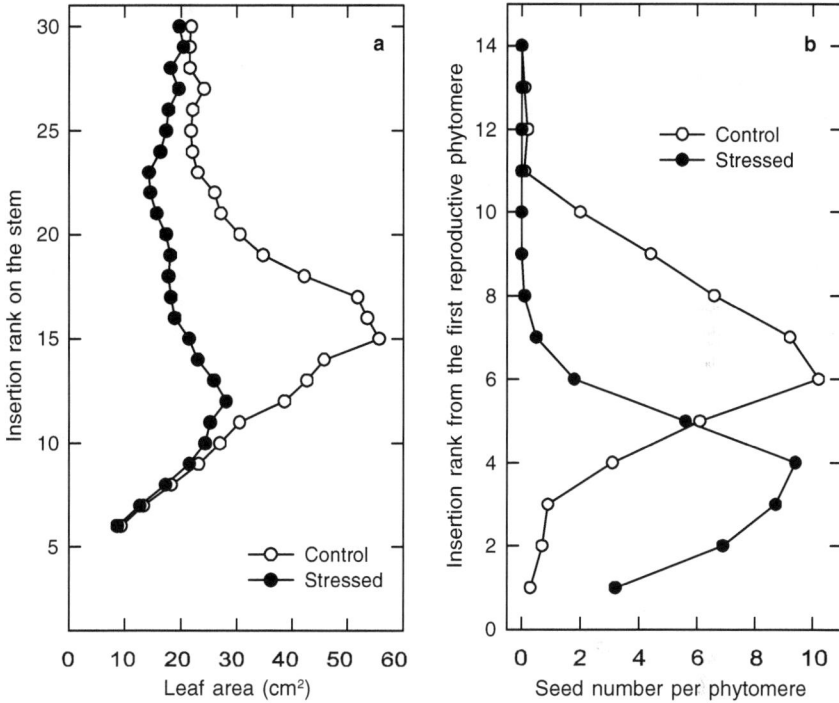

Figure 2.2. (a) Leaf area in relation to phytomere position on stem. Pre-flowering water deficit in greenhouse (see Lecoeur *et al.*, 1995) (b) Seed number per reproductive phytomer in relation to phytomere position from the first reproductive phytomer. Pre-flowering water deficit in the field (see Guilioni *et al.*, 2003).

and hydraulic root signals. However, the difference in diffusivity between water vapor and CO_2 allows a fall in CO_2 concentration of the sub-stomatic chamber, which occurs later than in transpiration. The second process relates to photosynthetic enzyme activity. It is reduced when a water deficit involves a fall in cell water concentration (water stress). This latter process corresponds to the deterioration of the quaternary structure of enzymes and photosystem (PSI and PSII), as well as to modifications in cell wall properties, primarily their permeability. These events occur when cells are severely dehydrated. They cause a fall in the production of reductive power at cellular level, but occur only in very severe water stresses.

Apart from situations of extreme water stress, the main process which limits photosynthesis in a pea canopy in the field is the closure of stomates which results in CO_2 supply reductions to the chloroplasts. The close relationship between the transpiration and photosynthesis responses to the water deficit results in quantitative relationships with the soil water state which take the same form for the two processes (fig. 2.1a and d) (Lecoeur and Sinclair, 1996; Guilioni *et al.*, 2003). In this way stomatic conductance and photosynthesis are maximal for FTSW greater than 0.4. Then, the two processes decrease in a linear manner to cancel themselves out when FTSW is exhausted. This effect of the water deficit is totally reversible in a few minutes when the plant is supplied with water again.

The changing radiation use efficiency in relation to soil water status is the same as for stomatic conductance and photosynthesis. This implies that the water deficit impacts on radiation use efficiency primarily through supply limitation in carbon chains and in reducing power for remaining metabolic activities. The other metabolic processes involved in building plant tissues are therefore relatively unaffected by water deficit. Proportionality between radiation use efficiency and soil water status for FTSW ranging between 0.4 and 0 makes it possible to carry out a simple assessment of the impact of a water deficit on biomass production. In severe and prolonged water stress conditions, biomass production can decrease by more than 80 %. In cases of moderate water constraint or water stress during reproductive development, reduction levels range between 20 % and 40 %.

Effect on yield and quality

• Number of seeds per plant

A water deficit which occurs between the beginning of flowering (BF) and the final stage in seed abortion (FSSA) reduces the number of seeds per plant (Guilioni *et al.*, 2003). The direct impact of a pre-flowering water deficit by limiting the number of fertilizable flowers or ovules is possible but occurs only in severe and prolonged water stresses.

The reduction in the number of seeds per plant is proportional to the intensity of the water deficit and results from the impact of the deficit on biomass production. Indeed, the relationship between plant growth rate between BF and FSSA is retained in any water deficit situation regardless of its intensity (Guilioni *et al.*, 2003). Thus, the impact of the water deficit on the number of seeds per plant does not result from specific direct effects on developing organs. The pea plant always overproduces seeds in embryogenesis. The number of filling seeds is then gradually adjusted to match the plant's capacity to supply assimilates between BF and FSSA. The growth rate of the plant during the seed number elaboration phase seems to be a good indicator of its capacity to fill seeds.

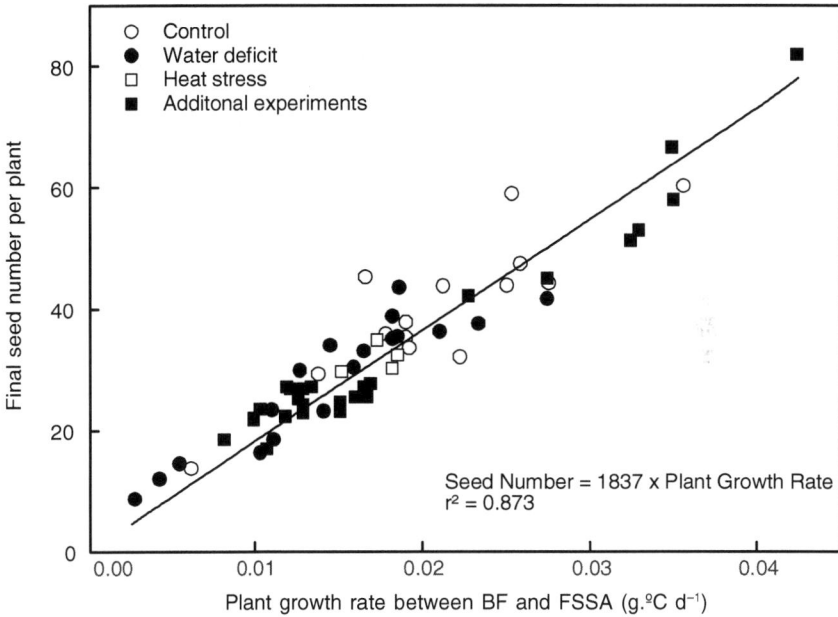

Figure 2.3. Relationship between the final seed number per plant and the plant growth rate between the beginning of flowering and the final stage in seed abortion. The straight line is the linear regression for the 4 data sets.

Some authors (Salter, 1962; Stocker, 1973) have shown the positive effects of a moderate water deficit on yield when occuring shorly before and after flowering. These effects could be explained by a limitation of vegetative development to the benefit of reproductive development (Turc et al., 1990) and a limitatition in seed abortions on the many first reproductive phytomers in non-limiting water conditions (Lecoeur, 1994a). These effects, however, which have been primarily observed in test networks or field experiments, could possibly result from an indirect improvement of sanitary status of the crop as a consequence of lower vegetative development.

• Distribution of seeds on different reproductive phytomers

A water deficit modifies the distribution of seeds along the stem (fig. 2.2b). In addition to reducing the final number of reproductive phytomers, it causes a concentration of seeds on the first reproductive phytomers (Guilioni et al., 2003). This concentration could be explained by a priority given to the first formed reproductive organs when assimilates production is limited. Furthermore, the impact of a water deficit on the expansion of young organs increases the advantage of the organs located on the first reproductive phytomers compared with those following. The plants experiencing water deficit present a larger number of seeds on the first reproductive phytomers in comparison to a well-watered plant. This excludes the hypothesis of direct effects of the water deficit on aborting seeds during their embryogenesis.

• Seed weights

Results on the impact of a water deficit on seed weight are contradictory. Certain authors report a depressive effect of the water deficit (Miller *et al.*, 1977; Harvey, 1980), an absence of effect (Ney *et al.*, 1994; Combaud, 1996) or a slight increase in seed weight (Turc *et al.*, 1990). These results show above all that seed weight is the result of biomass accumulation processes by the plant during reproductive development, the development of seed number, and finally biomass and nitrogen remobilisation during seed filling. Average seed weight can therefore be considered to be the expression of the phenotypical plasticity of the pea plant. Regulations within the plant during the reproductive development lead to a form of optimization of the number of filling seeds according to assimilate availabilities (carbon and nitrogen). Remobilisation of plant reserves is generally sufficient to maintain seed filling rate (Ney *et al.*, 1994), resulting in high stability of the seed filling rate except for situations of severe water deficits.

The main adjustment variable in yield development in response to a water deficit appears to be the number of seeds rather than seed weight. This result confirms limiting factor analyses in yield output on a field which emphasizes the predominance of seed number to explain the variability in the pea yield (Doré *et al.*, 1998). The majority of water deficits have, therefore, little or no effect on seed weight, except for severe deficits occurring before the end of the abortion limit stage for plants presenting a large number of seeds (> 2500 seeds.m^{-2}).

• Seed protein concentration

There is little information on the effects of water deficits on seed protein concentration. These effects could be primarily related to the impact of the water deficit on the nitrogen nutrition and the remobilisation of carbon and nitrogen reserves. Thus, it has been shown that, even in an optimal water supply situation, the nitrogen amount available for seed filling is almost always a limiting factor in seed protein concentration elaboration (cf. p. 123). Then, it is generally reported that these situations with optimal water supplies leads to a deterioration in seed quality, particularly in terms of protein concentration. A moderate deficit improves this quality (Turc *et al.*, 1990). On the other hand, a severe deficit leads to a fall in the content of the proteins as well as modifying their composition. There is, therefore, an optimal water supply pattern for seed quality, which involves controlling vegetative development, while not penalizing nitrogen nutrition.

- Germinative quality

A water deficit occuring during the seed filling period decreases the physiological quality of seeds, especially for early and intense water deficit. On the other hand, a water deficit occuring between the beginning of flowering and the beginning of seed filling is without effect on the germinative quality (Fougereux, 1994).

Conclusion

Water deficits have a wide range of effects. They affect all of the main plant functions: organogenesis and morphogenesis, photosynthesis and transpiration. The global response of a pea crop to a water deficit is thus as polymorphic as the water deficit itself, combining the position in the cycle, duration and intensity. Hence, it is difficult to give a single or average image of the impact of water deficit on yield development. In order to do so, we need a conceptual framework describing an outline of plant functioning at cycle scale, which identifies the primary phases and the important processes for each phase. This analytical framework then needs to be cross-matched with the probability of occurrence of water deficits which results from the interactions between the plant, the soil and environmental conditions. It is possible to carry out this exercise by using the energetic approach of biomass production proposed by Monteith (1977). By means of this approach, the pea cycle can be broken down into 3 main phases:

— a first phase installation of the leaves and roots from emergence to flowering during which plant leaf area and its maximal radiation interception efficiency are assessed;

— a phase of rapid biomass accumulation between BF and FSSA which depends primarily on the plant's capacity to transform radiation into biomass, hence its radiation use efficiency;

— a phase of remobilising reserves for seed filling between the beginning of seed filling and physiological maturity of the plant.

Armed with the knowledge of the most frequently observed water scenarios in the main French cropping areas it is possible to describe the stages of pea canopy functioning which are most frequently affected by water deficit.

The installation of the leaves or roots is seldom affected. Water extraction depth is rarely limited and the water deficit generally has little or no effect on changes in light interception efficiency from emergence until flowering. In this case the plant would need to experience an early and intense water deficit. The probability of this type of water deficit occurrence is low. On the other hand, final leaf area is frequently reduced. If this reduction generally does bring about a reduction in maximal interception efficiency at the beginning of reproductive development, it causes a more rapid fall in radiation interception efficiency at cycle end. This early fall results from a combination of a reduction in the number of reproductive phytomers and in the average area of the phytomers on top of the stem. For biological efficiency, low-intense and temporary falls can be observed shortly before flowering.

On the other hand, radiation use efficiency is very frequently reduced between BF and FSSA which causes a reduction in biomass accumulation rate and in the number of seeds per plant. The high probability of occurrence of these events, together with the large number of seeds in yield development reveals the period of the cycle between BF and FSSA to be the most sensitive to water deficit. Thus, it is recommended that the pea crop be irrigated during this phase. The frequency of water deficits during the remobilisation of plant reserves is even higher than during the preceding phase. However, the feedback which exists between biomass production and the number of filling seeds means that, in the majority of cases, seed weight is not limited.

Effects of high temperature on a pea crop

Lydie Guilioni, Marie-Hélène Jeuffroy

Short periods of high temperature have been shown to reduce pea yields. In the diagnosis of the yield-limiting factors, heat stress is often presented as a cause of yield variability between years, between regions and even within a region (Pumphrey et al., 1979; Jeuffroy, 1991, crop diagnosis UNIP-Arvalis-Institut du Végétal). The depressive effects of hot weather on yield are usually evaluated from the negative relationship between the final seed number and the cumulative temperature above 25°C, calculated during a critical period for seed determination, i.e. from the beginning of flowering to the final stage in seed abortion for the last seed-bearing phytomere. This agroclimatic criterion allows the identification of heat stress as a factor responsible for localised differences between potential and actual yield.

Nevertheless, it is not good enough to assess the specific impact of heat stress on each step of yield elaboration and to understand the effects of high temperature on crop physiology.

Numerous studies have dealt with the effects of high temperature on a particular aspect of crop physiology or on various yield components. The aim of this chapter was not to synthesize all these results but to analyze the effects of heat stress at the plant lifecycle level. Yield elaboration was divided into 3 stages: a first phase of plant leaf area establishment, a second phase of rapid biomass accumulation from the beginning of flowering and the last phase of leaf senescence linked to seed filling. The effects of heat stress on yield were investigated, taking into account the stress intensity and its timing in relation to these 3 phases.

In a first part, we characterised the heat stress experienced by a pea crop in agronomic conditions. Then, we successively investigated the effects on leaf area and radiation interception, on rate of biomass accumulation, on plant lifespan and seed number. Results were partly based on experiments with various sowing dates (Guilioni, 1997; Guilioni et al., 1998) making it possible to obtain similar heat stresses at different times in the plant cycle.

Characterization of heat stresses

A constraint is defined by its intensity, duration and position in the plant life cycle.

• Air and plant temperature

Air temperature is usually used to describe plant temperature. But the air temperature at the top of the crop canopy level may be different from the temperature sensed by the plant in the field and even in a controlled environment. Measurements of the temperature of different organs of a pea plant (apical buds, floral buds, pods) have shown thermal gradients inside the canopy (Guilioni, 1997). Generally, temperatures of the organs located in the apical part of the plant and subjected to solar radiation (apical buds, young leaves and flowers) were similar. Pods inside the canopy receiving less radiation were a few degrees cooler than flowers or young reproductive organs located in the apical bud (fig. 2.4). So they were less subject to heat stress. Apical bud temperature was higher by 2 to 4°C than air temperature at standard height (1,5 m above the soil surface) during the day and lower at night (fig. 2.4). Characterization of heat stress by air temperature alone, as is usually the case in studies of heat stress, may lead to a poor estimate of the heat stress experienced by plants. It is noteworthy that during the day, apical buds were cooler than surrounding air (fig. 2.4), even in growth chamber experiments (Guilioni *et al.*, 1997). Consequently, controlled experiments generally overestimate the stress experienced by plants.

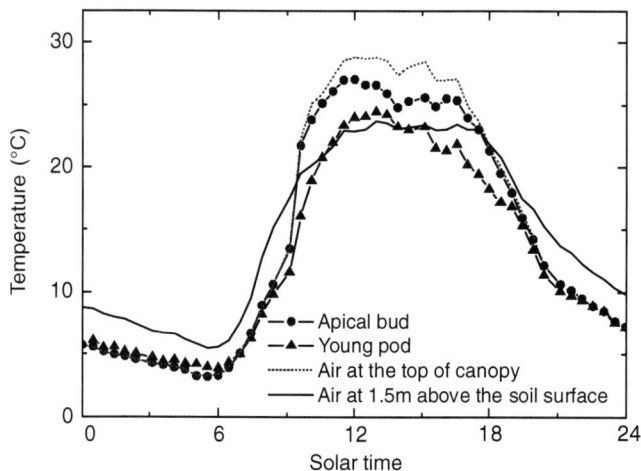

Figure 2.4. Example of diurnal variation of air and plant temperatures (from Guilioni, 1997). During the day, air temperature differs significantly from plant temperature, and plant temperature decreases with depth inside the canopy.

The risk of heat stress increases with water shortage. When soil water deficit leads to stomatal closure and reduces plant transpiration, plant temperature increases (Lecoeur and Guilioni, 1998; fig. 2.5). In the case of severe water deficit, the difference between well-irrigated and drought-stressed plants may reach several degrees (fig. 2.5c).

• Definition of a threshold of temperature for heat stress

For a pea crop, a temperature above 25°C can be considered as limiting. For example, photosynthesis decreases for temperatures above this value (Guilioni *et al.*, 2003). Although it is quite easy to define an optimal or limiting temperature for a basic process such as an enzymatic activity, it is much more difficult for an integrative process like growth or yield elaboration. Regarding the differing effects of temperature on the plant's physiology in each temperature range, stress-inducing temperature has been defined as moderately severe when plant temperature exceeds 30°C for a few hours a day for a few days, or severe if it exceeds 35°C for similar periods.

• Frequency of heat stresses in France

A frequency analysis of air temperatures during the period of seed formation was conducted in four areas of pea production in France (west,

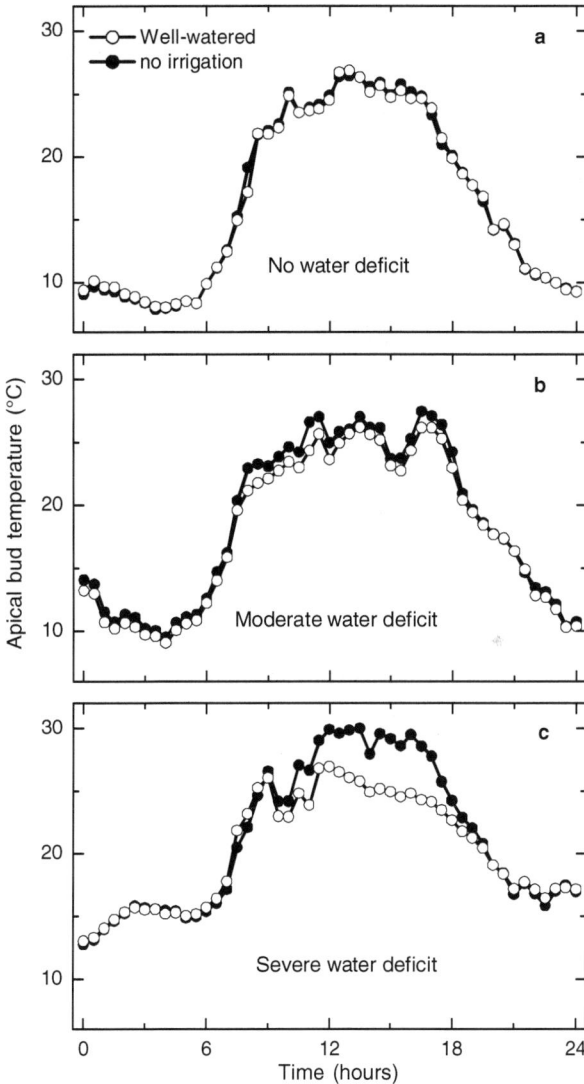

Figure 2.5. Diurnal variation of apical bud temperatures of irrigated or water-limited pea plants (from Lecoeur and Guilioni, 1998). When plant transpiration is reduced by water deficit, plant temperature increases by several degrees during the day.

Parisian Basin, south-west and Rhone Valley) for weather records taken over 21 years. It was shown that during the period of yield elaboration, a pea crop might experience a few days of moderate heat stress (Guilioni, 1997; Guilioni *et al.*, 1998). The probability of encountering heat stress increased with the time of sowing. Conversely, severe heat stresses (T>35°C) were very rare and generally occurred late in the plant life cycle, when the final seed number is already determined. It is noteworthy that pea crops are less subjected to heat stress in the south of France, although the climate of the south is hotter than in the north. In fact, high temperature appears essentially in June and July whatever the region, but in the south a pea crop reaches the final stage in seed abortion in June, even for late sowings. So, heat stresses come too late to harm yield.

Heat stress effects

- Various effects depending of the timing of stress in the plant cycle

The effects of high temperature depend on the period of application. Similar heat stresses occurring at different periods in the plant life cycle have different effects on yield. For crops with no water or nitrogen limitation, experiencing heat stress at various periods in the plant life, the earlier and more persistent is the stress, the more yield and above-ground biomass are reduced (Guilioni et al., 1998; fig. 2.6a and 2.6b).

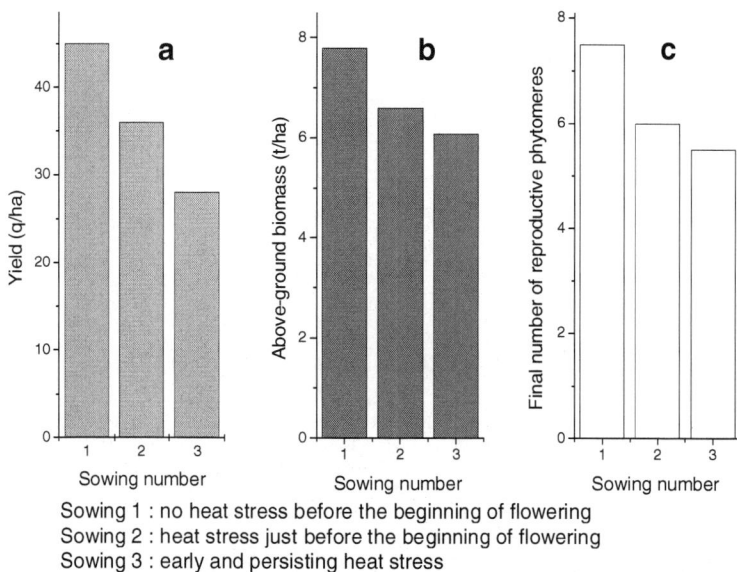

Sowing 1 : no heat stress before the beginning of flowering
Sowing 2 : heat stress just before the beginning of flowering
Sowing 3 : early and persisting heat stress

Figure 2.6. Effects of heat stress position on seed yield (a), on dry above-ground biomass (b) and on final number of reproductive phytomeres (c). Field experiments, cultivar Messire (from Guilioni et al., 1998). Sowing 1: daily mean air temperature (Tair) under 18°C; sowing 2:7 days between the floral initiation and the beginning of flowering with Tair>23°C; sowing 3: heat stress similar to sowing 2 but experienced at emergence and Tair>20°C during the entire crop cycle. The earlier and more persistent the stress, the more yield, above-ground biomass and final reproductive phytomeres are reduced.

- Effects of heat stress on establishment of leaf area and on radiation interception

Over the range of temperatures experienced by pea crops in France, high temperatures do not affect the rate of leaf appearance (Guilioni, 1997). However they do reduce the size of all the leaves growing at the time of the stress, i.e. the ten or so young leaves located within the apical bud.

The more severe the stress, the greater is the reduction observed. Nevertheless, only a very severe reduction in total leaf area will impair the radiation interception and therefore the biomass production. In fact, a leaf area index of 3 is optimum for radiation interception: the radiation interception efficiency (RIE) is more than 80 %. A leaf area index greater than 3 increases self-shading but does not significantly improve RIE. When heat stress occurs just before the beginning of flowering, the decrease in leaf area has no effect on RIE (fig. 2.7b). Plants reach an optimal leaf area index despite the reduction in size of some leaves. Conversely, a heat stress at the beginning of growth may reduce the leaf area of the first phytomeres, leading to a substantial decrease in RIE (fig. 2.7b). So, according to the frequency analysis conducted for the various areas of pea production in France, reduction of RIE by heat stress is very rare. It might be observed only in the case of late sowing dates with high temperature occurring just after emergence and well before flowering.

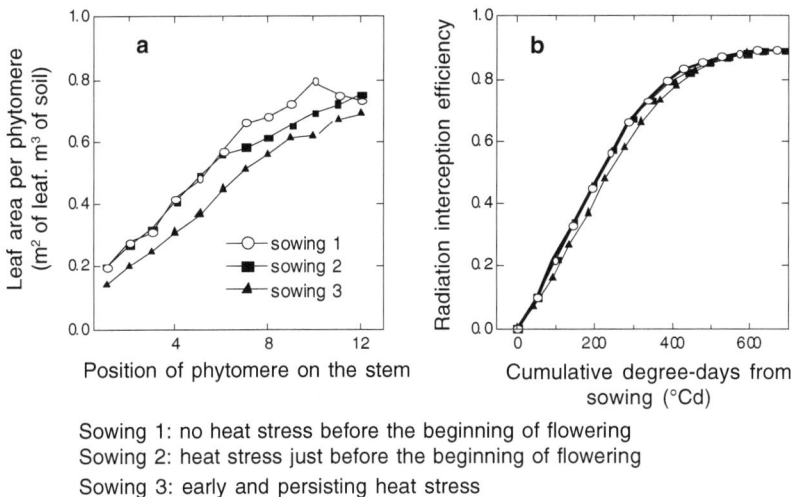

Sowing 1: no heat stress before the beginning of flowering
Sowing 2: heat stress just before the beginning of flowering
Sowing 3: early and persisting heat stress

Figure 2.7. Effects of a heat stress on the leaf area per phytomere (a) and on the radiation interception efficiency (b) (from Guilioni *et al.*, 1998). Any heat stress reduces the size of the developing leaves but only a very early heat stress may reduce the radiation interception by the crop.

• Effects on the rate of biomass accumulation

A high yield requires a leaf area large enough to insure a good radiation interception combined with a high capacity to convert intercepted radiation into biomass (cf. part "Carbon acquisition at the crop level in pea"). The conversion efficiency is called radiation use efficiency (RUE). This process is especially important from the beginning of flowering when the rate of biomass accumulation is maximal (Lecoeur and Ney, 2003). In fact, the final seed number per plant is linearly related to the plant growth rate between the beginning of flowering and the final stage in seed abortion (Guilioni *et al.*, 2003; p. 105). RUE hardly depends on the plant net photosynthesis (Pn), i.e. gross photosynthesis minus respiration.

In peas, the maximum net photosynthesis was measured when leaf temperature was 15–27°C. From 28–40°C, Pn gradually decreased (fig. 2.8, Guilioni *et al.*, 2003). Severe heat stress reduced Pn by as much 33 %. Whatever the timing of stress in the plant's life, a heat stress, even short and moderate, reduces RUE via a decrease in Pn. The rate of biomass accumulation is affected (fig. 2.6b) which may lead to a yield decrease (fig. 2.6a).

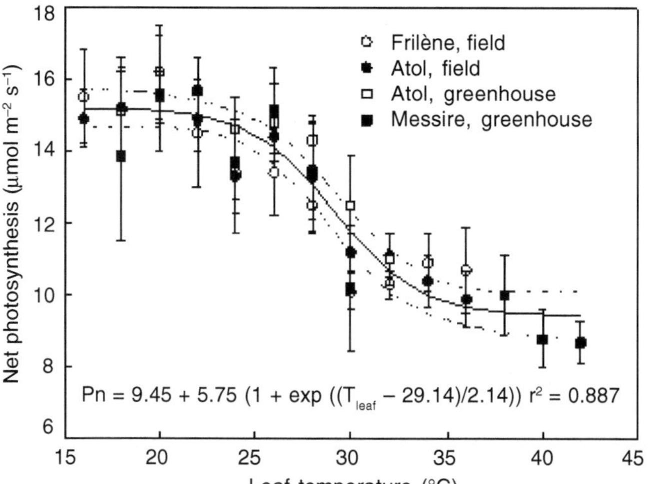
Frilène, field
Atol, field
Atol, greenhouse
Messire, greenhouse
Pn = 9.45 + 5.75 (1 + exp ((T_leaf – 29.14)/2.14)) r² = 0.887
Net photosynthesis (µmol m⁻² s⁻¹)
Leaf temperature (°C)
18
16
14
12
10
8
6
15
20
25
30
35
40
45

Figure 2.8. Net photosynthesis of a pea leaf as a function of leaf temperature (from Guilioni *et al.*, 2003). Mean and confidence intervals at P=0.05. The continuous line is the common fit for all the cultivars and dotted lines indicate the confidence interval of the fit at P=0.05. Net photosynthesis is maximum between 20 and 25°C and decreases at higher temperatures.

- Effects on the plant lifespan

The major effect of a moderate heat stress, as is most common French pea-growing areas, is a decrease in the final number of reproductive phytomeres (Stanfield, 1966; Guilioni et al., 2003). The earlier and more persistent the stress, the greater is the reduction (Guilioni et al., 1998; fig. 2.6c). A period of high temperature changes the trophic balance between the vegetative organs growing in the apical bud and the reproductive organs (floral buds, developing embryos, filling seeds). The reduction in the growth of vegetative organs would decrease their assimilates demand. Hence, reproductive organs would become more competitive compared to vegetative organs. This might accelerate the normal cessation of phytomere production by the apical meristem and thus shorten the plant lifespan. Conversely, in cases in which moderate heat stress is applied when the plant has already ceased production of new phytomeres (as in Jeuffroy's experiments (1991) where only two more leaves were produced after the heat stress), the final number of reproductive node is not affected.

As often observed in field pea, a short period of mild heat stress may reduce the number of flowering phytomeres. Nevertheless, it is not an abrupt cessation of flowering during the period of heat stress. It happens several days or even a few weeks later. Flower abortions observed in this case are similar to these observed in the absence of heat stress, when apical senescence occurs. They occur on the upper part of the plant, near the last expanded leaf (Guilioni et al., 1997). The effect of mild heat stress should be viewed as the acceleration of a programme linked to the normal termination of phytomere production during the plant cycle and not as an abrupt event linked to stress.

Heat stress may decrease the plant lifespan reducing the final number of phytomers. A severe heat stress may have the opposite effect (Guilioni et al., 1997) if the stress is early and severe enough (T>35°C) to stop the development of floral buds immediately and to cause their rapid abscission. In this case, the final number of phytomeres is increased in direct proportion to the number of flowering nodes with floral abscission (Guilioni et al., 1997) and the lifespan is increased. Such conditions are rarely experienced by pea crops in France.

- Effects on seed number

It is not possible to analyze at the phytomere level the effects of heat stress on the final seed number, except when severe heat stresses impair a step in seed development, so a mild heat stress before the beginning of flowering may decrease the number of seeds on some phytomeres and increase it on others (Guilioni, 1997). The potential reduction is not an immediate consequence of the stress on the development of the floral buds and embryos. The effects of heat stress on seed number must be analyzed at the plant level, taking into account the indeterminate growth pattern of pea and the relationships between the different organs on the plant.

The decrease in seed number depends on the intensity, duration and timing of the stress within the life cycle. Thus, for a brief moderate heat stress, the number of phytomeres may be decreased without reduction in the seed number. The number of seeds is increased on the basal part of the plant (Guilioni, 1997). When the stress is strong enough to limit assimilates availability, it leads to abortion of some seeds which have not yet reached the final stage in seed abortion (Jeuffroy, 1991). Peas have a high reproductive plasticity, adjusting the potential number of reproductive sinks in an apparent balance with the availability of assimilate in the plant. So, despite the complex relationships between the different organs of the plant, heat stress does not affect the linear relationship between seed number and plant growth rate between the beginning of flowering and the final stage in seed abortion (Guilioni *et al.*, 2003). Rather than directly impairing particular steps in gametogenesis, fertilisation or embryogenesis, heat stress seems to act by affecting carbon metabolism and the rate of biomass production.

Conclusion

The pea crop physiology is disturbed whenever air temperature exceeds 25°C, i.e. for daily mean temperature greater than 20°C. This temperature is frequently observed during the life of a pea crop in France. Rather than a rare weather event, heat stress should be regarded as one of the limiting factors for pea production in France.

Studying the effects of heat stress at the organ level allows the identification of stages in yield elaboration susceptible to high temperature, but does not allow the prediction of the effect on yield. For example, a moderate late heat stress may decrease or increase the number of seeds per phytomere, depending on the phytomer concerned (Guilioni, 1997). The effects of heat stress have to be analyzed at the plant level.

The energetic basis of biomass production proposed by Monteith (1977), associated with a frequency analysis of the climatic risks in the French areas of pea production, allowed the most commonly affected mechanisms to be identified. The more frequent stresses are moderate ones (temperatures below 35°C). Such heat stresses decrease the radiation use efficiency, *via* the photosynthesis, and if the stress occurs early in plant life reduce the final number of phytomeres and the lifespan. These two effects contribute to a yield decrease linked to a reduction in the rate and duration of biomass production. Moderate heat stresses also reduce the area of the leaves developing during the stress, with no effect on the radiation interception efficiency. Heat stress leading to a decrease in the radiation interception efficiency are exceptional. They may occur early in plant life, between emergence and floral initiation.

It is noteworthy than heat stress effects are similar to those observed for water deficit: decrease in leaf area, in photosynthesis and shortening of the lifespan. Water deficit and hot weather are often associated. Water shortage leads to an increase in plant temperature. So the effects of water deficit and heat stress are additive, and cause substantial yield reduction.

Nitrogen deficiency

Aurélie Vocanson, Marie-Hélène Jeuffroy, Thierry Doré

In several species, nitrogen deficiencies can have detrimental effects on the crop performances, sometimes leading to high yield losses: wheat (Demotes-Mainard et al., 1999; Jeuffroy and Bouchard, 1999), maize (Plénet and Cruz, 1997), oilseed-rape (Colnenne et al., 2002), forage crops (Bélanger et al., 1992). A limiting nitrogen nutrition of the crop can lead to a reduced leaf area (Bélanger et al., 1992; Plénet and Cruz, 1997) and then to a reduction of the amount of intercepted radiation by the crop, of the photosynthesis and of the radiation use efficiency in biomass (Sinclair and Horie, 1989; Sinclair and Muchow, 1999), and finally of the biomass production (Bélanger et al., 1992). On these species, relationships have been proposed to quantify the effect of nitrogen deficiencies on these variables. The relationships are stable if the nitrogen deficiency is characterized by the nitrogen nutrition index, which allows to determine the precise period of deficiency and its intensity.

Characterization of a nitrogen deficiency

• Nitrogen nutrition index

In order to give a good account of the nitrogen nutrition status of a crop and to identify some periods of deficiency, Lemaire and Gastal (1997) proposed to calculate the nitrogen nutrition index (NNI), which is the ratio between the observed nitrogen content of the aerial parts of the crop (%Nobs) and the critical nitrogen content, calculated from the aerial biomass of the crop (%Nc), at each stage during the crop cycle, as defined by the critical dilution curve (cf. p. 61–62):

$$NNI = \%Nobs\ /\ \%Nc$$

This indicator allows:

- To determine the period of occurrence of a nitrogen deficiency, when NNI is significantly lower than 1;
- To quantify the intensity of the nitrogen deficiency : the lower the NNI, the more intense the deficiency. If the NNI is higher than 1, the crop growth is not limited by nitrogen and, if no other limiting factor exist, is that allowed by the instantaneous climate.

- ## Nitrogen deficiencies exist on a pea crop

The pea plants have the possibility to fix atmospheric nitrogen by the symbiosis with the bacteria of *Rhizobium leguminosarum,* largely present in most soils in the world (Amarger 1991; Catroux 1991), and also to assimilate mineral nitrogen as most of the other grown species. Numerous authors have shown that these two ways of nitrogen uptake were complementary (cf. § I.3.2). At the beginning of the cycle, when the nodules are not formed on the root system, the uptake of mineral nitrogen is often the main way. Then, its importance decreases with the presence and the increase of the nodule activity. In the soils characterized with a high availability of mineral nitrogen, the N_2 fixation is generally limited, while the uptake of mineral nitrogen is high. These observations led some authors to assume that one way of nitrogen uptake always compensated for the other, therefore assuming that a pea crop could never be affected by nitrogen deficiencies. Yet, nitrogen deficiencies have been frequently observed in farmers'fields grown with pea (Doré, 1992; Crozat, *et al.,* 1994).

A precise diagnosis of the functioning of a crop characterized by a nitrogen deficiency shows that the N deficiencies can be the result of a water stress, of a limiting P or K nutrition, of a compacted soil structure (of the seedbed and/ or the ploughed layer) or of insect damage by *Sitona lineatus* L. All these factors affect nodule formation, with consequences on nitrogen fixation (Doré, 1992). The reduction of mineral nitrogen uptake by roots can also be linked to the presence of diseases, as *Aphanomyces euteiches* (Engqvist, 2001), to a root system little developed, consequence of a compacted soil structure, for example (cf. p. 166), or to a low mineral nitrogen availability in the soil (low mineralisation, water stress, ….). In these conditions, the mineral nitrogen uptake does not allow to compensate for the low N_2 fixation, leading to a nitrogen deficiency. The environment and the crop management have then a high influence on the nitrogen nutrition of the crop (Wéry, 1987).

Effects on plant development

Although Truong and Duthion (1993) have observed a reduction of the duration of the period between plant emergence and beginning of flowering when the plant nitrogen content is low, the date of beginning of flowering is, from most authors, little affected by the level of N nutrition in the field (Sagan, 1993), as in controlled conditions (Jeuffroy,1991).

In fact, a nitrogen deficiency even early in the crop cycle does not seem to have an influence on the position of the first reproductive node: the reproductive initiation occurs around the stage 5–6 leaves, this stage generally occurring before the first visible consequences of a nitrogen deficiency. On the other hand, a reduction of the flowering duration was observed either after an early nitrogen deficiency, occurring before the beginning of flowering and lasting during the flowering period, or for a low level of nitrogen nutrition from plant emergence (Jeuffroy and Sebillotte, 1997; Roche *et al.*, 1998). In these cases characterized by a nitrogen deficiency, the number of reproductive nodes is reduced and can be limiting of the seed number and finally of the crop yield.

A nitrogen deficiency can also limit the rate of leaf emergence but does not seem to have an effect on the rates of reproductive stages progression along the stems, flowering and beginning of seed filling (Sagan *et al.*, 1993; Roche *et al.*, 1998). The date of beginning of seed filling at the stem scale could however be earlier due to a nitrogen deficiency (Ney, comm. Pers.).

Finally the number of branches existing at the beginning of flowering also varies when the level of nitrogen nutrition varies : when the nitrogen nutrition is high at the beginning of flowering, there are numerous branches, for a given range of plant densities (Doré *et al.*, 1998).

Effect on the plant aerial growth

For non-fixing mutants with a low N fertilisation, the plant aerial growth is severely reduced when a nitrogen deficiency occurs (fig. 2.9; Sagan, 1993).

For experiments with non-nodulating mutants and classical cultivars receiving various levels of nitrogen fertilisation, the reduction of biomass linked with the nitrogen deficiency can be calculated as follows (fig. 2.10):

$$DM/DMmax = 1.11 \times (1 - 3.35 \exp (-5.80 \times NNI)) \ (r^2=0.88)$$

With DM/DMmax: ratio between the dry matter of the deficient treatment and the dry matter of a treatment without N deficiency at the time t; NNI of the deficient treatment at the same time t.

The amount of dry matter produced every day depends on the amount of radiation which is intercepted by the crop (mainly depending on the global radiation, on the leaf stand and on the leaf area of the crop) and on the radiation use efficiency, (Monteith, 1977; cf. p. 44). In maize, Plénet and Cruz (1997) showed that a nitrogen deficiency essentially affects RUE. In fact, the risks of nitrogen deficiency are higher at the end of the cycle, period characterized by a high N requirement (from the filling seeds) and

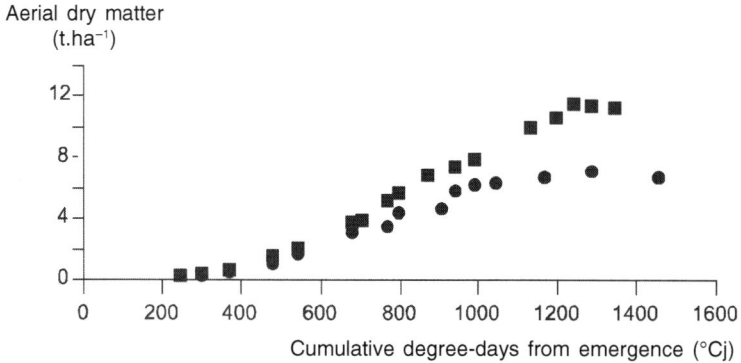

Figure 2.9. Aerial growth curves for a non-nodulating mutant non-fertilized (●) and a non-nodulating mutant fertilized with 250 kg N.ha⁻¹at sowing (■) in relation to sum of degree-days from emergence (Dijon, 1997).

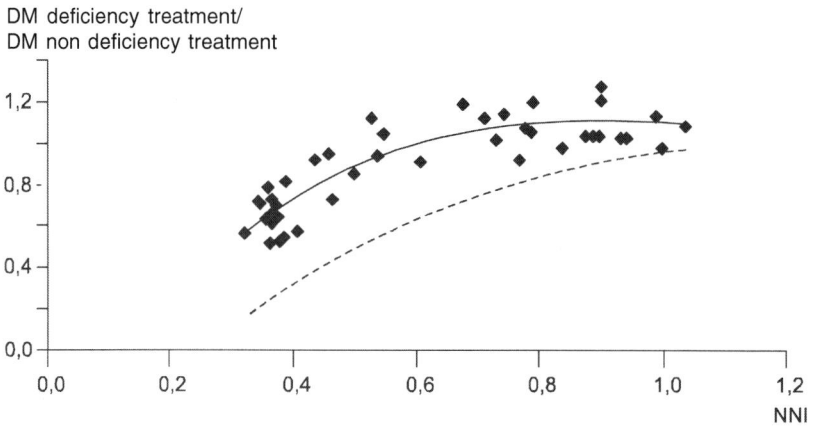

Figure 2.10. Loss of aerial dry matter (on deficiency treatment/on non-deficiency treatment) in relation to NNI of the treatment (data=INRA Dijon, 1997, 2000). (♦ = experimental points used to parameter the relation; — = relation for pea; --- = relation for rescue).

a severe progressive reduction of N availability at plant scale. At this period, the crop has reached its maximum leaf area. The reduction of the nitrogen availability in the plant leads to a decrease of the photosynthesis and then of ε_b. Besides the effect of the lower leaf nitrogen content on ε_b, a nitrogen deficiency can also lead to a reduction of leaf area of the plant, also explaining the lower plant growth.

On fig. 2.10 is shown the comparison between the effect of a nitrogen deficiency on plant growth on a pea crop and on a fescue crop. A pea crop seems to be less sensitive to a nitrogen deficiency than a fescue crop. This relationship should be verified on a larger database, but a possible explanation of the difference between the two species could be linked to the difference of metabolic efficiency, similar to that suggested by Plénet and Cruz (1997) in the comparison between maize and sorghum.

Effect on seed yield and quality

Seed yield and seed protein concentration are reduced when the pea crop is characterized by a nitrogen deficiency (Sagan, 1993). The main yield component affected by a nitrogen deficiency is the seed number per m² (Crozat et al., 1992), the mean weight per seed being little affected. Jeuffroy (1991) showed that the seed number is more reduced when the period of nitrogen deficiency occurs early in comparison with the date of beginning of flowering and has a long duration. A nitrogen deficiency after the beginning of flowering seems to have little effect on this component. These results are consistent with the high correlation showed by Doré et al. (1998) between NNI at the beginning of flowering and the seed number per m². The reduction of seed number could be linked to the simultaneous effect of the nitrogen deficiency on the reproductive nodes (Jeuffroy and Sebillotte, 1997) and on the crop growth rate. Finally, the effect of a nitrogen deficiency on the production variables is highly linked to its duration and its position in the crop cycle. Therefore the crop nitrogen concentration and the crop biomass at the beginning of flowering appear as an interesting tool to predict the potential seed number per m² at this stage (Crozat et al., 1992).

A nitrogen deficiency before the beginning of flowering leads to a lower seed protein concentration (Crozat et al., 1992). The nitrogen deficiency affecting numerous process of the plant (growth, nitrogen uptake, nitrogen remobilization from the vegetative organs to the seeds, …) and finally the amounts of nitrogen in each organ.

Conclusion

A nitrogen deficiency can occur on a pea crop when N_2 fixation and/or mineral nitrogen uptake are limited by various factors such as the presence of disease, a compacted soil structure, a water stress, the presence of *Sitona lineatus* L., … A long or intense nitrogen deficiency leads to the reduction of biomass production on the plant, and detrimental effects on seed number and yield, but also on seed protein concentration. The NNI at the beginning of flowering is correlated to the branching of the plant, determinant of the potential seed number per m². Thus the NNI seems to be a pertinent tool to forecast the effects of a nitrogen deficiency in a pea crop.

Effects of compacted soil structure

Aurélie Vocanson, Yves Crozat, Marie-Hélène Jeuffroy

In farmers' fields, a compacted soil is a frequent cause of limiting yields (Doré, 1992). Unfavourable soil structures are mainly linked with soil tillage after ploughing and/or sowings in too wet conditions. These conditions are frequent in humid years because there are few available days to work in the fields in the early spring, particularly in the farms where several species (sugar beets, spring barley, pea, faba beans, ...) must be sown during a short period and other technical operations have to be done (nitrogen fertilisation on cereals, herbicides application, ...).

The available data concerning the effects of soil structure on a pea crop presented in this chapter come mainly from experiments with a comparison of contrasted soil structures, the sowing bed being always optimal in order not to modify plant emergence (Crozat et al., 1991; Crozat et al., 1992). The apparent density of the ploughed layer, but also the proportion of compacted clods with low internal porosity ("delta clods") (Manichon and Gautronneau, 1987), are used to characterized the soil structures observed.

Effects on plant development

No modification of plant development during the vegetative period was observed in case of compacted soil structure. The beginning of flowering can sometimes be earlier and the duration of flowering is always shorter, from 4 to 8 days. The rates of progression of flowering and SLA along the pea stem are not modified. The consequence is then an earlier date of maturity of the crop.

Compacted soil structures can also have detrimental effects on the number of branches on the plant (Jeuffroy and Ney, 1997; Doré et al., 1998). This effect is probably due to a reduction of crop growth.

Effects on crop aerial growth

The difference of plant biomass accumulation between compacted and non compacted situations are rarely significant at the beginning of flowering. Afterwards, these differences increase until harvest, with an earlier growth stop in severely compacted soils. When the proportion of compacted clods in the ploughed layer is higher than 70 %, a decrease in biomass production can be observed (fig. 2.11): in the loamy soils where these results were obtained, the biomass loss ranges from 20 % to 30 %, for proportions of compacted clods around 70 % and can reach 60 % for a more compacted soil structure. The effects of a compacted soil on aerial growth can be partly the consequence of the negative impact of the compaction on the ability of the crop to extract water and minerals from the soil (Tardieu and Manichon, 1987).

In case of a water and/or thermal stress, the effect of a soil compaction on the biomass production is lower (fig. 2.11: case of Grignon and Mons, 2003): the reduction of biomass production is around 10 % for a proportion of delta clods of 70 %. Whatever the soil structure is compacted or not, a water stress limits the yield potential reducing the effects of the compacted soil structure on yield (Crozat *et al.*, 1991).

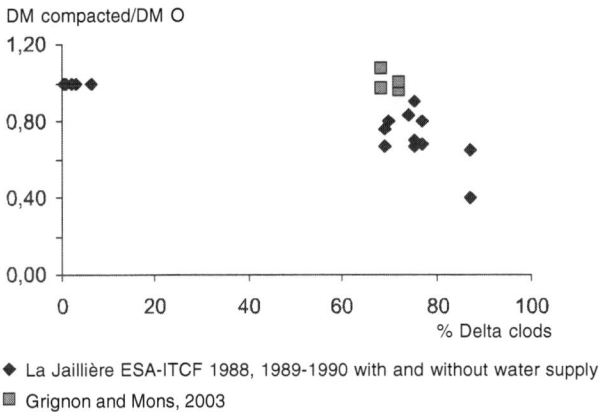

♦ La Jaillière ESA-ITCF 1988, 1989-1990 with and without water supply

▩ Grignon and Mons, 2003

Figure 2.11. Relative loss of total aerial dry matter (DM) (DM of compacted treatment B–C/DM of noncompacted treatment (= control) O) at harvest in relation to proportion of delta clods (% Delta clods) in the ploughed layer.

Effects on the root system and on the nodulating system

• Effect on the root depth

The maximal root depth is reduced at the 6–7 leaves stage if the ploughed layer is compacted (table 2.1). At this stage, the root depth in a compacted soil is half of that of the control, which is then just under the ploughed layer. Afterwards, the difference of root system between the two treatments is maintained all along the crop cycle, indicating that the progression rate of maximal root depth is the same in the soil layer located under the compacted zone in the two situations (table 2.1). Thus, the compaction of the ploughed layer leads to an early and stable reduction of the progression rate of the rooting depth. The rooting rate can be reduced until 20 % in compacted soils, with a maximal root depth decrease of 20 %. Similar observations were realised in autumn sowings (fig. 2.12).

The root progression is divided in 3 distinct phases:

- • phase 1 : slow progression of the maximal root depth corresponding to the winter period. In compacted soil, the root progression is limited by the resistance of the ploughed layer,

Table 2.1. Effect of compaction (compacted ploughed layer (C) compared to an non compacted ploughed layer (O)) on the progression of the root depth (cm) (Trials La Jaillière ESA-ITCF 1988, 1989, 1990 irrigated treatments).

	Stage 6–7 leaves			Stage FFSA (maximal root depth)		
	O	C	Difference O-C	O	C	Difference O-C
1988	34	18	16	65	37	28
1989	32	17	15	79	56	23
1990	41	20	21	64	48	16
mean	36	18	17	69	47	22

FFSA = Final Stage in Seed abortion

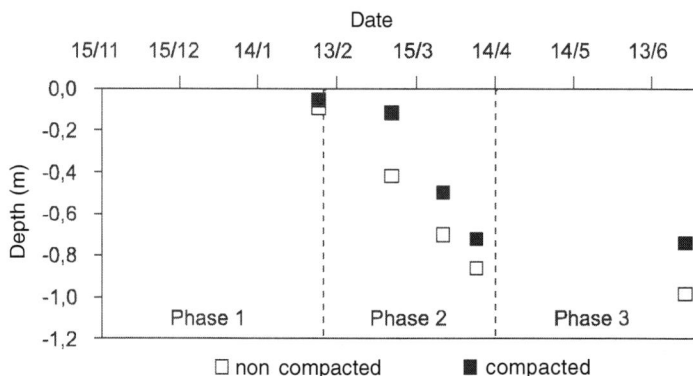

Figure 2.12. Comparison of rooting note for a compacted treatment and a non-compacted treatment (11/15/02 sowing, Dove variety, Grignon).

- phase 2 : rapid linear growth. The rate of root depth depends on the obstacles, particularly of the delta clods encountered by the roots (Tardieu and Manichon, 1987),
- phase 3 : growth end.

In the case shown on figure 2.12, the compacted soil structure decreased the maximal root depth from 25 %. To day, no significant difference between varieties was observed concerning the effect of a compacted soil structure.

• Effect on the root density

Another effect of the compacted soil structure was observed on the root density. However, the modalities of root distribution depend on the crop stage. For an early stage in a compacted soil (Vocanson, np), roots are more frequent in the sowing bed (0–10 cm), the compacted ploughed layer preventing them to grow in the soil. Afterwards, fewer roots are observed in the ploughed layer of a compacted soil than in a non compacted soil (fig. 2.13), probably because a higher mechanical resistance of the soil (Tardieu and Manichon, 1987).

Root density—compacted/control

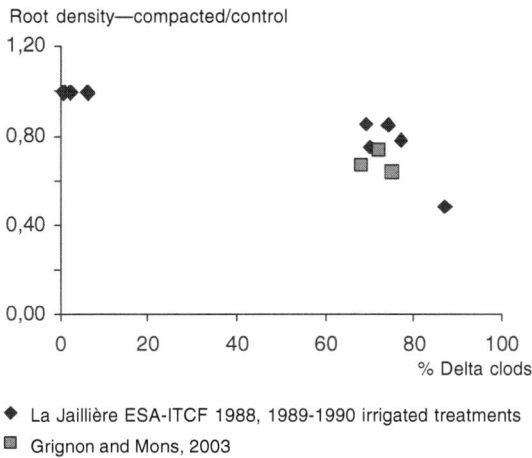

● La Jaillière ESA-ITCF 1988, 1989-1990 irrigated treatments
▫ Grignon and Mons, 2003

Figure 2.13. Relative index of root exploration in the ploughed layer (% of cases filled by treatment B and C/% of cases filled by control O) at FFSA stage in relation to proportion of Delta clods (% Delta clods) in the ploughed layer.

• Effect on the nodules

The compaction of the ploughed layer does not affect the total number and the total biomass of the nodules, but it modifies their spatial partitioning (table 2.2). In the compacted zone, the lengthening of the taproot is reduced and nodules are mainly located in the basal part of the taproot. A high decrease of the number of nodules is then observed. This reduction is compensated for by an increase of the number of nodules on the lateral shallow roots, located in the non compacted sowing bed. Therefore if the soil is compacted, the nodules are nearly all located in the sowing bed where the porosity is more favourable.

The compaction seems to have a negative direct effect on the nodules at the taproot scale, but, at the scale of the whole root system, compensations could exist, which could explain the absence of effect of the compaction on the size of the whole fixing system. This compensation process confirms the hypothesis of a regulation of the nodulation by the aerial parts of the plant (Pierce and Bauer, 1983; Kosslack and Bohlool, 1984), either by a genetic determinism (put into evidence by the diversity among pea varieties: varieties without nodules, called non-nodulating varieties, or presenting a high number of nodules, called hyper-nodulating (Caetano-Anolles and Gresshoff, 1990; Sagan and Gresshoff, 1996), or by trophic regulation of the size of the fixing system (Bethlenfalways *et al.*, 1978; Tricot *et al.*, 1997).

Table 2.2. Effects of compaction of the ploughed layer on some characteristics of the fixating system of pea (Tricot, 1993) ; BF = beginning of flowering, 6L = 6 leaves.

	O (control)	C (compacted)
Number of nodules per plant at BF	137 (± 55)	188 (± 30)
Number of nodules of the taproot per plant at BF	39 (± 10)	19 (± 8)
Depth of the last nodule of the taproot at stage 6L (cm)	15.7 (± 5.2)	9.2 (± 3.2)
Biomass of nodules at BF (mg-per plant)	55 (±8)	52 (±9)
Part of the biomass located in the sowing bed (%)	82	97

Few knowledge is available on the consequences of a change in the nodule partitioning in a soil profile. For a compacted soil structure, Crozat *et al.* (1991, 1992) observed an early senescence of the nodules and the rapid stop of the fixing activity, according to the drying of the surface soil layer. On the opposite, in a non compacted soil, the nodules located deep in the soil were active during a longer period.

Effects on the nitrogen nutrition

On the whole data of the experiment Arvalis-Institut du Végétal-ESA, the amount of nitrogen accumulated in the crop at harvest in a compacted soil was decreased by 37 % in comparison with that observed in a non compacted soil (Crozat *et al.*, 1991,1992), which is in the same range than the relative loss of accumulated biomass. At the beginning of the seed filling period, the nitrogen nutrition index (p. 157–158) varies from 0,58 to 0,67 for the compacted soil structures. This reduction of the nitrogen nutrition is linked with a significant reduction of the amount of fixed N per ha (–45 %) and of the amount of assimilated mineral (–41 %).

This reduction of mineral uptake could be due to a lower mineralisation of the organic nitrogen in compacted soil (Jensen, 1993) and/or to a lower ability of the crop to uptake the available nitrogen. This latter could be linked with differences in root system or with changes in the water transport due to compaction, and then to a lower accessibility of the roots to the various nutrients (Tardieu *et al.*, 1987). Several elements are in favour of the hypothesis of the lower root density. On one hand, a deficient P and K crop nutrition is also observed in compacted soils (Crozat *et al.*, 1992). On the other hand, compaction has no or few effect on the %Nitrogen derived from symbiotic fixation (Crozat *et al.*, 1992), while this proportion is highly sensitive to a variation of the nitrate content of the soil area with roots, in absence of water stress (Voisin, 2002). In these compacted soils, the reduction of the amount of fixed nitrogen could be the consequence of the reduction of aerial growth (particularly after the beginning of flowering), reducing the assimilate availability required for the nodules activity, and not a direct consequence of the compaction. Moreover, an early stop of fixation could be linked to the drying of the surface soil layer.

Effects on seed yield and quality

An average yield loss of 18 % is observed in compacted soils on the experiments realised by Arvalis-Institut du Végétal. However, the differences between treatments can be highly variable.

- In the case of water stress and/or of high temperature during the flowering period, the negative effect of a compaction is hardly visible, even sometimes null. As for the biomass increase, this can be due to a drying of the soil surface layers as rapid in non compacted soils than in compacted soils. Consequently, the advantage of the non compacted soils in terms of amount of available water is not visible.
- When the potential yield of the control is higher, higher yield losses can be observed (50 %). In this case, the yield loss can be explained by a reduction of the seed number, linked with a lower number of reproductive nodes (Crozat *et al.*, 1994). This effect of the compaction on seed number can be linked with a low nitrogen nutrition index at the beginning of flowering in a compacted zone. The mean weight per seed is little affected by the compaction : this confirms the lower variability of this component comparatively to that of the seed number per m² (Jeuffroy and Ney, 1997).

If the compaction does not always lead to a yield loss, the seed protein concentration is always reduced from 1 % to 3 % in the crops grown on compacted soils. For a given year, the nitrogen nutrition index in compacted soils is always lower than that of the control and, on all years, we can observe a positive linear relationship between the NNI at the beginning of seed filling and the seed protein concentration (Crozat *et al.*, 1994).

Conclusion

The compaction of the ploughed layer has a direct effect on the rooting process, limiting the progression of the root depth, the root density and the maximal root depth. In compacted soils, nodules are mainly located in the first 10 cm of the soil, making them more sensitive to soil drying near the surface. The soil compaction also affects the water nutrition and the mineral nutrition of the crop, with consequences on the biomass produced and the amount of nitrogen cumulated. Finally, a compacted soil structure decrease seed yield (through a reduction of the seed number) and the seed protein concentration. However, the effects of the compaction depend on the water nutrition conditions and on the temperatures at the end of the growth cycle.

A part of the yield variability in pea can then be explained by the soil structure. In average, an autumn sowing is realised in better water soil conditions than a spring sowing (Meynard, 1985), the new cultivars of winter pea under breeding (type Hr; p. 16 to 18), by a larger range of possible sowing dates (from the beginning of october), could limit the risks of soil compaction at sowing.

Cold temperatures and the functioning of the canopy in pea

Isabelle Lejeune-Hénaut, Bruno Delbreil, Rosemonde Devaux, Lydie Guilioni

Like wheat, rape and lupine, cultivating winter peas has been envisaged with the aim of increasing production potential and ensuring regular yield. And it is true that sowing in autumn enables growth cycles to be structured differently over the year (fig. 2.14). Firstly, the cycle begins when temperatures are generally lower, which enables the cycle to be extended by several weeks. This means the crop can intercept the solar radiation over a longer period and, in theory, produce more biomass (Monteith, 1977), and in particular a

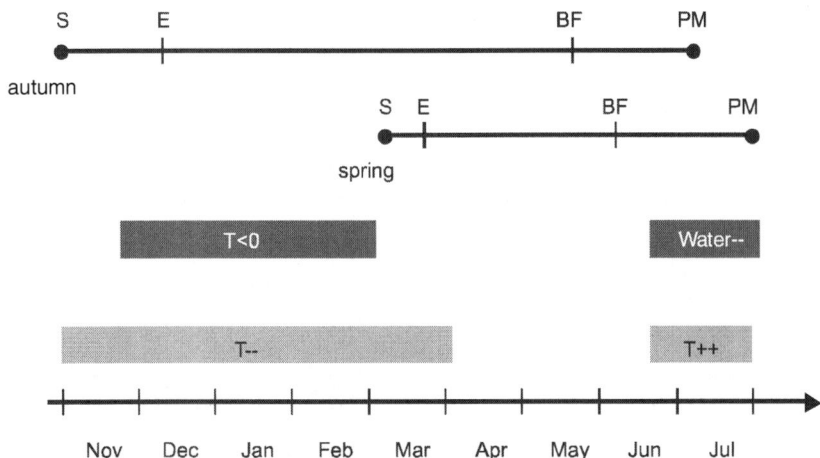

Figure 2.14. Schematic diagram of key developmental stages and periods of major climatic risk for field peas sown in autumn or spring in the North of France.

Sowing (S), emergence (E), beginning of flowering (BF), physiological maturity (PM), thermal constraint: low temperatures (T --) or high temperatures (T ++) or frost (T <0), water deficit (Water --).

higher seed yield (fig. 2.15). Secondly, the cycle ends earlier in the year making it possible to avoid unfavourable climatic conditions at this stage (water deficit, high temperatures). In practice however, there are certain limits. When peas are sown in autumn, the canopy develops in sub-optimal thermal conditions (low but not freezing temperatures), and to survive the winter, has to withstand periods of frost that can vary in length and in intensity depending on the region. Frost can cause damage that ranges from partial tissue necrosis to death of the whole plant (Etévé, 1985; Lejeune-Hénaut and Wéry, 1994). Furthermore, the often wetter climatic conditions in winter favour the development of fungal diseases such as *Aschochyta* blight (Tivoli & Samson, 1996), considerably limiting the development of winter pea crops in northern latitudes.

In the first part of this chapter we explain what this thermal constraint (low temperatures) implies in pea. We then review the known effects of

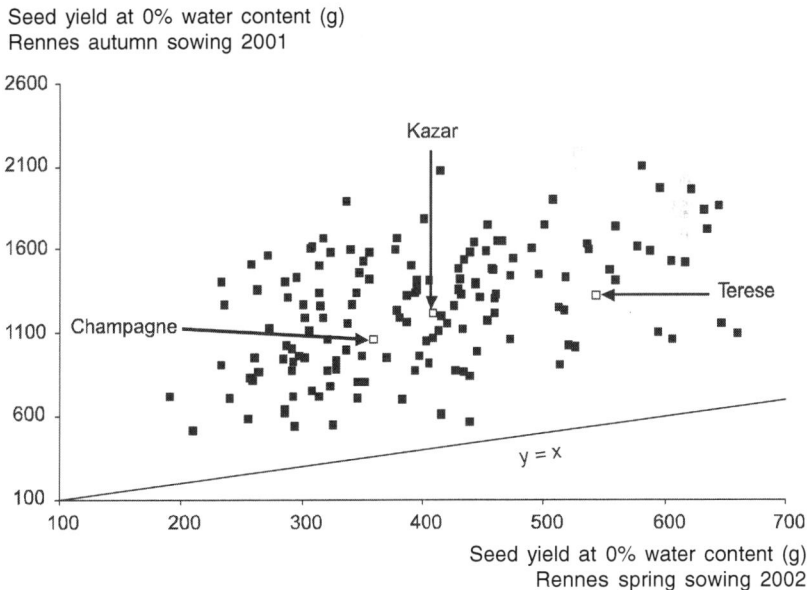

Figure 2.15. Comparison of seed yields obtained with spring and autumn sowings in 144 recombinant inbred lines derived from a Champagne x Térèse cross.

The seed yields correspond to the weight of seeds harvested on plants grown on wire (i.e. approximately 55 plants). Lines are indicated by filled squares (□), controls by empty squares (■), and the bisector by a solid line.

The autumn sowing trial was realised on October 30, 2001, and the spring sowing on March 26, 2002. Maximum fungicide protection was applied (2 and 3 treatments in the autumn and spring sowings respectively) to limit the development of *Ascochyta* blight. Winter temperatures in 2001–2002 were relatively mild (5 days of negative mean temperatures with a minimum of –7°C) which enabled us to record yields for both sowing dates.

sub-optimal temperatures and of frost in pea. Finally, we show how a period of acclimation, generally under low temperature and short day conditions, can result in hardening, i.e. enable the plant to become more resistant to frost.

Definition of cold as a thermal constraint

The extent of damage caused by cold depends on the severity (minimum temperature), and the duration of the constraint, but also on the stage of development of plant organs—which display different degrees of sensitivity to cold—and on the history of the crop, which may or may not favour cold acclimation. A constraint is thus defined by its intensity, duration and timing in the cycle of the crop.

• Air temperature and the plant

The characterization of this type of constraint within a given plot is typically based on the temperature of the air measured under cover at a height of 2 metres from the ground. However, this temperature is not entirely representative of the temperature of the plant. For example, if the canopy is not fully developed, the soil plays an active role in exchanges of energy between the plant and its environment. During the day, the apices of the peas are then generally warmer than the surrounding air and during the night they are colder (fig. 2.16) with differences in temperature of up to 6°C. These differences vary depending on climatic conditions and are smaller when the sky is cloudy, but there is no simple relationship between the temperature of the air and the temperature of the plant. Only modelling the energy balance of the vegetation enables these differences to be calculated (Lhomme and Guilioni, 2004).

• Definition of a cold-stress temperature threshold

It is possible to distinguish between a constraint (the period at which the temperature is not optimal) and a stress, which has more serious consequences for the functioning of the canopy. In all cases, it is essential to identify the temperature threshold below which plant functioning is modified or symptoms appear. Defining optimum temperature is extremely complex when dealing with integrative processes like development or the elaboration of yield.

When symptoms are used to define the threshold of sensitivity to cold, the stage of development of the crop has to be taken into account, and this has already been accomplished in certain species such as fruit trees. In pea,

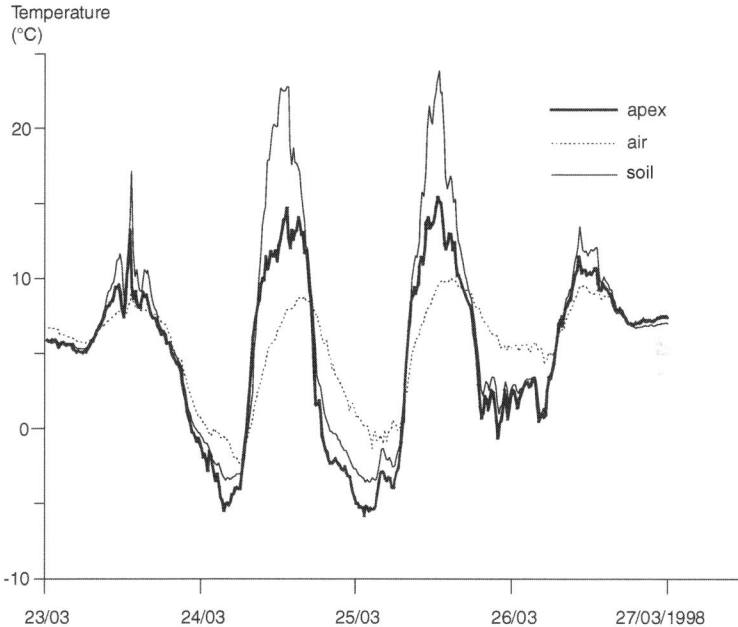

Figure 2.16. Example of daily variations in air temperature at a height of 2 m above the ground, in temperature at the soil surface and in temperature of apical buds, in a plot of winter peas in Grignon (78, France). Cheyenne variety sown on October 28, 1997 (Guilioni, unpublished data).

some results can be considered as references for lethal temperatures as a function of a given developmental stage of the plant. For example, observations on the variety Cheyenne (fig. 2.17) showed that the degree of resistance depends on the stage of development of the plant: resistance, expressed as a percentage of plants that died, appeared to be maximal at the two-leaf stage, became much less uncertain at the four-leaf stage and had practically disappeared at the six-leaf stage, which corresponds to floral initiation in this variety. According to observations made during the winters of 2001–2002, 2002–2003, and 2003–2004, the threshold of resistance of the vegetative apex of Cheyenne after cold acclimation is around –15°C.

Other studies indicate that genetic variability exists in pea with regards to resistance to frost. Among fodder peas, some genotypes can withstand minimum temperatures of around –15°C after a preliminary period of hardening (Lejeune-Hénaut, unpublished data). Our current knowledge of the ability of pea to survive frost, although still fragmentary, enabled us to propose a schematic diagram of survival curves at low temperatures (fig. 2.18).

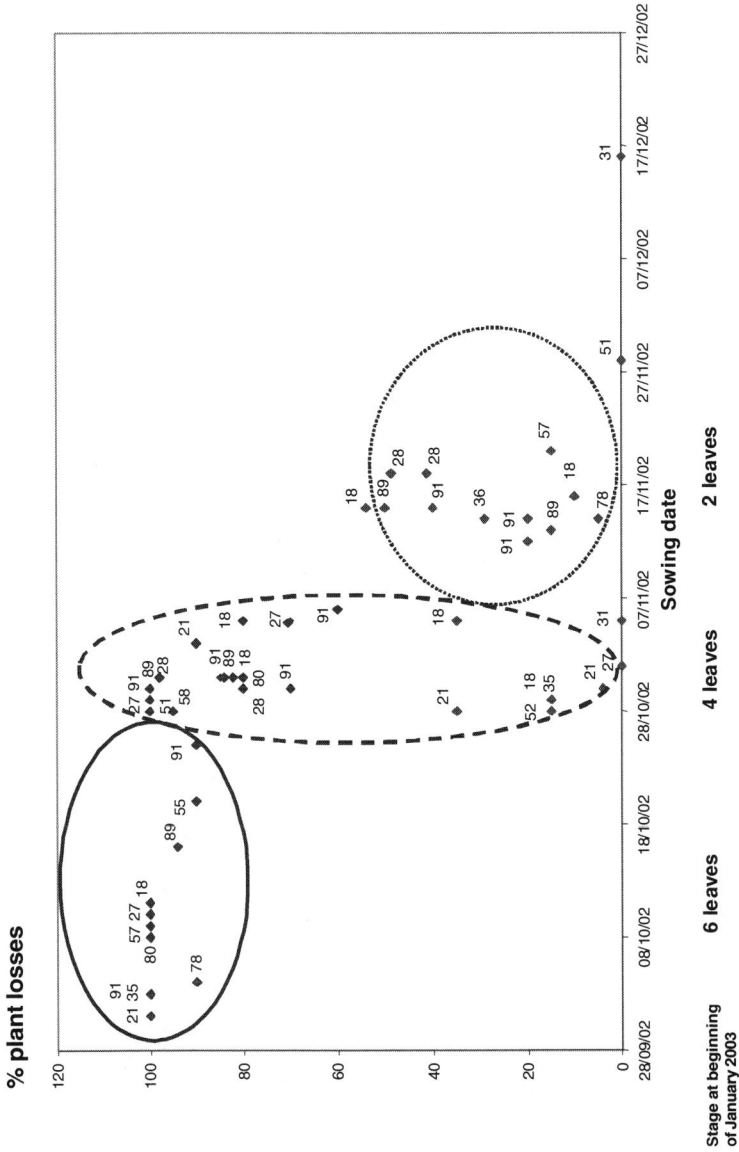

Figure 2.17. Influence of the developmental stage on the percentage of plant losses observed during the winter 2002–2003, for cv. Cheyenne. Numeral labels of the points on the graph correspond to the administrative division of France where the field was located. (From Biarnès, Perspectives Agricoles n°293, September 2003).

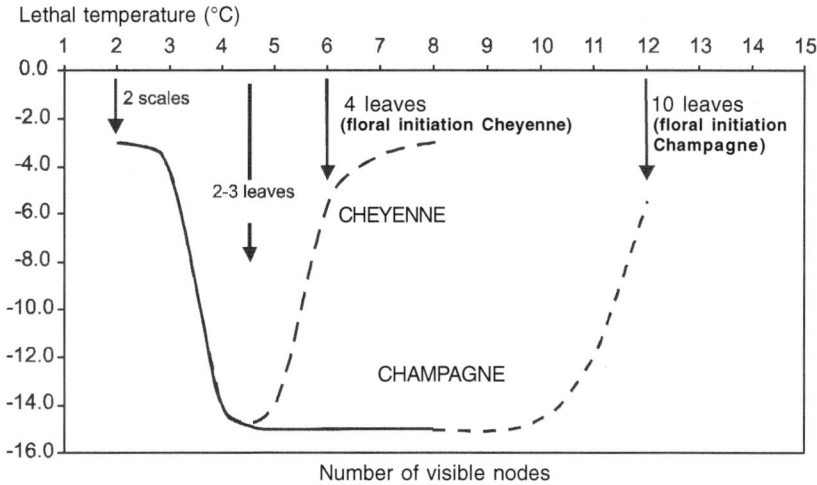

Figure 2.18. Schematic diagramme based on field observations for the two lines. Only the dashed line for Champagne is hypothetical, since no frost was observed at the time of floral initiation (mid-April) in this genotype (Lejeune-Hénaut, Biarnès & Chaillet, unpublished data).

- ## Cold temperatures in France

Cold and sub-optimal temperatures are often experienced by pea canopies, particularly by crops sown in autumn. For example, frequency analysis of the risk of frost or cold temperatures for a pea crop sown in November in different regions in northern France showed that air temperatures below –5°C occured for 3 to 16 days along the period with cumulated negative temperatures on a –5°C basis reaching –38°C (Du Montant, 2001). In this respect, the North-Eastern region of France is the least favourable for pea during this period. For a variety like Cheyenne, the risk of temperatures below –15°C occurring between the two-leaf stage and the eight-leaf stage is very low and even zero eight years out of 10. On the other hand, between the eight-leaf stage and the beginning of flowering, mean temperatures below 10°C that can limit the production of biomass (p. 58–59) are very frequent and can occur during two thirds or more of this period of development.

Effect of cold temperatures on the functioning of the canopy

Before describing damage caused by frost, we review the effects of low positive temperatures on plant functioning.

• Effect of low positive temperatures

To our knowledge, few studies have been performed on the specific effects of cold on pea. The studies that have been conducted focussed either on physiological functions such as photosynthesis, or gave a description of symptoms at the scale of the plant or of the canopy. Concerning photosynthesis, Feierabend et al., (1992), reported photoinhibition of photosystem II at temperatures below 15°C. Yordanov et al., (1996) and Georgieva and Lichtenhalter (1999) demonstrated the reversible effect of treatment at 4°C on photosynthetic activity with, in particular, a reduction in the transport of electrons at the level of photosystem II.

In relation with the reduction in photosynthesis, cold has a depressing effect on the radiation use efficiency (RUE). Based on data in the literature, and on experiments carried out on staggered sowing dates, a relationship has been proposed between RUE and mean air temperature, where RUE is expressed relatively to the maximum RUE observed at a mean air temperature of 18°C (Guilioni, 1997, see also p. 58–59). RUE is reduced at air temperatures of below 12°C. Thus peas sown in the autumn are exposed to temperatures that are sub-optimal for radiation use efficiency during almost the entire vegetative stage. This could explain why, even in the absence of frost, the performance of winter peas often fails to reach its full potential. The relation between RUE and temperature within the range of cold temperatures was consequently studied in more detail using data from a multi-site pea network comprising several different organisations (Montpellier SupAgro, ARVALIS-Institut du végétal, INRA, UNIP) in 2001 and 2002 (Guilioni, unpublished data).

Peas cultivated in France are never exposed to air temperatures below 2°C after flowering. However, studies conducted in Australia (Brand et al., 2003) showed that episodes of air temperatures below 2°C during seed filling result in serious losses in yield (up to 35 %). Many seeds (up to 60 %) also displayed discolouration of the seed-coat resulting in a reduction in quality. Consequently, when varieties are grown for human consumption, down-grading the product to animal feed involves serious financial loss.

• Frost damage

At the level of the cell

Frost generally causes the formation of crystals whose presence has varying repercussions on the cells (Mazliak, 1992):

— when the temperature drops slowly, crystals form on the outside of the cells, i.e. in the intercellular spaces and on the surface of the organs; their presence results in a reduction in the water content of the cells (water stress); this is accompanied by an increase in the concentration of solutes in the cell environment (osmotic stress); the elimination of water causes a reduction in cell volume and distortion of the cell wall (mechanical stress). If the plasma membrane and other cell membranes can withstand the entry and exit of water that accompanies the freeze/thaw cycles, the cell can survive;

— when the temperature drops rapidly, the quantity of water lost by the cells is not sufficient to avoid intra-cellular crystallization; the plasma membrane tears (mechanical damage) rapidly causing the death of the cell.

At the level of the organs and of the whole plant

Not all the plant tissues and organs are equally sensitive to frost. Underground organs are generally less tolerant than above-ground organs. The resistance of buds varies as a function of their age: the least resistant buds are those located in apical position on the stems (Dereuddre and Gazeau, 1992). In pea, Etévé (1985) described morphological differences in the terminal apex before and after floral initiation. Reproductive buds are bigger than vegetative buds due to the larger cells of the floral primordia. Microscope observations showed that during a period of frost, the water in the meristem cells is evacuated preferentially towards the vegetative organs at the base of the bud that are not yet fully differentiated. These observations provide a hypothesis to explain the increased sensitivity of apices to frost after floral initiation. In fact, if one considers that the survival of the terminal bud depends on the speed of evacuation of water towards the young leaves at its base, a floral meristem will have proportionally more water to evacuate than a vegetative meristem, and will consequently be more sensitive.

When cells are destroyed by frost, damage that is visible to the naked eye can range from partial necrosis of tissues to death of the whole plant. In agronomic conditions, damage is usually evaluated at the scale of the whole plant (see score card for frost damage p. 180). The first symptoms are burns at the borders of the leaf blades followed by progressive necrosis from the top to the bottom of the plant. The survival of the individual plant depends on the proportion of the stem that remains undamaged at the base and on the condition of the root system: ideally, axillary buds located lower down the stem can take over from the terminal meristem when the latter is destroyed. For the last 30 years, classical selection has been based on this overall evaluation of plant survival and has enabled partial improvement of resistance to frost in dry winter peas. Nevertheless, additional improvement would be possible if the genetic determinism of resistance to frost was identified more precisely, on the basis of biochemical and physiological mechanisms called into play during a decrease in temperature. This is the aspect we discuss in the following paragraph.

Adaptation of pea to winter conditions

In plants, low positive temperatures are the most common inductor of cold acclimation. In the autumn, the drop in temperatures results in changes in many biochemical and physiological processes in the plant. Among the changes observed, those that represent the response called cold acclimation result in an increase in resistance to low positive temperatures and frost, and to other winter stresses (anaerobic stresses caused by flooding or ice, dehydration caused by wind, photoinhibition caused by the combination of cold temperatures and high solar radiation). In addition to a decrease in temperature, a reduction in daylength is often required for cold acclimation. The latter is important in many woody species and in a number of herbaceous species. In pea, the reduction in photoperiod induces a response at the developmental level: in the case of autumn sowing, the plant remains longer in the vegetative state enabling it to avoid freezing. The genetic control of the ability to avoid freezing is described p. 16 to 18. It is mainly ensured by the *Hr* gene which blocks transition to floral initiation when the days are short. This locus has been shown to play a significant part in the variability of freezing resistance, together with two main other areas (i.e., QTL, Quantitative Trait Loci) in the pea genome (Lejeure-Hénaut *et al.*, 2008). We now consider in more detail the acclimation mechanisms that occur before floral initiation is achieved and lead to cold acclimation.

• Evaluation of cold acclimation ability

A review of the literature (Pearce, 1999) shows that the main physiological and biochemical changes that take place during acclimation are:

— activation of primary metabolism (respiration, photosynthesis, protein synthesis);
— accumulation of solutes (sugars, free amino acids, etc.);
— changes in the composition of membrane lipids, which help maintain fluidity of cell membranes when temperatures are falling;
— activation of protective systems against oxydative stresses (production of ascorbic acid, transformation of glutathione into its reduced form, etc.).

Several studies on pea varieties with very different levels of resistance to frost (Bourion *et al.*, 2003) reported variations specifically linked to acclimation, both with respect to the dry matter content of the plant, and to levels of cryoprotective molecules.

• Accumulation and distribution of dry matter in different organs

Pea plants accumulate more dry matter per degree-day when the plant is subjected to cold acclimation temperatures (mean daily temperature of around 5°C; fig, 2.19) than when they develop at temperatures that are *a priori* more favourable (mean daily temperature of around 15°C). These observations have frequently been reported in the literature for annual herbaceous plants, both cultivated species like wheat and rape (Hurry *et al.*, 1995) and the model species *Arabidopsis thaliana* (Wanner and Juntilla, 1999). In pea, whatever the level of frost resistance of the variety tested, this acclimation effect is more marked in roots than in above-ground organs. However, varietal differences do exist in the amount of dry matter accumulated above ground.

Thus, before the acclimation phase, young Térèse seedlings, a sensitive dry variety, have the advantage over Champagne seedlings in terms of root dry matter, probably linked with the heavier weight of the individual seed. However, this advantage disappears during cold acclimation, and, even if the two varieties displayed the same absolute values for root dry matter at

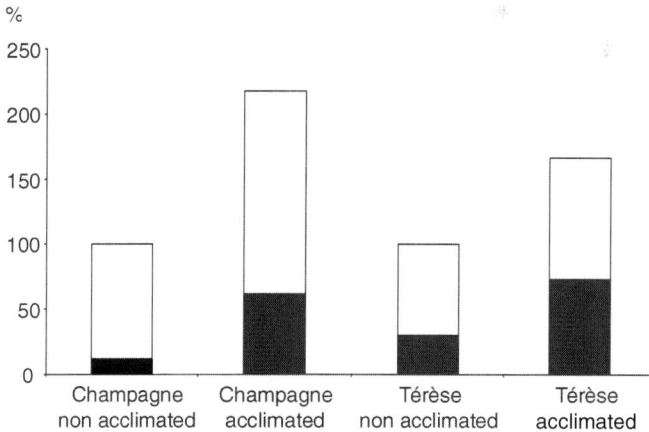

Figure 2.19. Proportion of above-ground (white bars) and underground (black bars) dry matter in the varieties Champagne and Térèse after 11 days in nursery conditions (19°C/ 12°C day/night, photoperiod: 10 hours, light intensity: 250μmoles photons m^{-2} s^{-1}), followed by:
- acclimated: 10 days at 10°C/2°C day/night;
- non-acclimated: 4 additional days in nursery conditions.

All the plants had 3 true leaves when sampled. Dry weight was measured on 5 plants with 3 repetitions for each variety and each condition, acclimated or non-acclimated. For a given variety, the total area of the histogram is proportional to the total dry weight measured in non-acclimated plants, i.e. 66 and 102 mg/plant for Champagne and Térèse respectively.

the end of the experiment, only cold acclimated Champagne seedlings survived frost after the cold acclimation period. Figure 2.19 shows that accumulation of dry matter during hardening was higher in the resistant variety Champagne than in Térèse, and that the root system was the main benificiary of the increase. The increase in the allocation of resources to undergound organs has already been reported in perennial plants like alfalfa (Boyce and Volenc, 1992; Castonguay et al., 1995) and white clover (Guinchard et al., 1998). However, there is not always a direct correlation with varietial differences in frost resistance; which is why these authors were also interested in the exact distribution of the different types of carbohydrates.

- • Accumulation of different carbon compounds during cold acclimation

According to the literature, variations of certain carbohydrates during the course of cold acclimation appear to be correlated with the acquisition of frost resistance. A QTL for raffinose concentration during cold acclimation in the field was observed on both linkage groups (LG) 5 and 6. Both these QTL colocalize with a QTL for frost resistance (Dumont et al., 2009). Raffinose is a soluble sugar which has been shown to have a major effect during cold acclimation for different species. In alfalfa, another legume species, accumulation of raffinose and stachyose has been positively related to frost resistance (Castonguay et al., 1995). Even if the role of raffinose in cold acclimation and freezing resistance is still debated in Arabidopsis (Taji et al., 2002; Zuther et al., 2004), our results strongly suggest its implication in frost resistance of pea. Additionally, a QTL for the RuBisCO (Ribulose 1,5-biphosphate carboxylase/oxydase) activity colocalizes on LG6 with a QTL for frost resistance. It supports the hypothesis that this key enzyme of the Calvin cycle, that allows fixation of CO_2 during the photosynthesis, also plays an important part in pea frost resistance. To further illustrate the importance of the carbohydrate metabolism, the 2-D analysis of proteins differentially expressed during cold acclimation allowed to detect PQL (Protein Quantitative Loci) related to carbohydrate metabolism. Particularly, PQL for triose phosphate isomerase and fructose bisphosphate aldolase, which are involved in glycolysis together with PQL for RuBisCO small subunit and chaperonin 21 which are implied in photosynthesis, were found to colocalize with the QTL for frost resistance on LG5 (Dumont et al., 2009).

Several different functions have been proposed for sugars in their role of protecting cells against frost damage: they could act as cryoprotectors of certain enzymes (Carpenter et al., 1986), or be involved in stabilizing cell membranes during the movement of water in successive stages of dehydration and rehydration (Lineberger & Steponkus, 1980; Gusta et al.,

1996; Danyluk *et al.*, 1998), or act as osmolytes to prevent excessive dehydration of the cells during frost and during the formation of intracellular ice (Steponkus, 1984). From data recorded in pea, (Bourion *et al.*, 2003) reported that concentrations of soluble sugars did not appear to be high enough to induce noticeable variations in osmotic pressure during the course of cold acclimation, but these results remain to be confirmed. Similar observations were made by Guinchard *et al.* (1997) who showed that in white clover only 30 to 40 % of variation in osmotic potential during acclimation could be explained by the concentration of soluble sugars. The cryoprotective function of raffinose mentioned by Gusta *et al.* (1996) is thus of particular interest in pea.

Conclusion and outlook

The central role played by carbohydrates in the process of acclimation to cold makes their different functions of primary concern in research on resistance to cold and to frost. Our understanding of the signalling network in which they are integrated is increasing very rapidly in *Arabidopsis thaliana* (Ruelland *et al.*, 2009) and could benefit pea, provided that pertinent characteristics can be measured in this agronomically interesting species. Thus, in pea, the first step will be to identify the origin of the accumulation of dry matter and in particular of carbohydrates, during acclimation. Levitt (1980) proposed a schematic diagramm of the adaptation of carbon metabolism to cold that deserves particular attention. At 25°C, the majority of products of photosynthesis are used for growth and respiration, whereas at 5°C, growth is almost completely halted but respiration is only slowed down, resulting in higher accumulation of carbohydrates. This statement could be valid for pea as it has been observed that the resistant variety Champagne displays a specific morphology in autumn: the leaves and internodes are very small and the plant does not start active growth until the following spring. Smaller organs have quite often been observed during acclimation of frost resistant plants (Hurry *et al.*, 1995; Li, 1994).

In another connection, it is very important to control the pleiotropic effects of QTL allelic variation on the productivity of the population. To achive this objective, the contribution of ecophysiological models is indispensable as they will enable identification—during progressive introgression of useful genes into new varieties—of factors that limit the production of seeds. In particular, generalization of a model to determine the rate of survival of a pea population as a function of mimimun temperatures at the location concerned, the stage of development of the plant, and conditions of cold acclimation, will be necessary to predict the biomass produced by a canopy of peas sown in the autumn. To build this model, precise lethality/survival curves will be needed that integrate available genetic variability for minimum temperature thresholds for survival and the speed of cold acclimation of the plant. The use of models would also enable simultaneous integration of allelic variations for genes potentially involved in cold acclimation, and for those that determine freezing avoidance (Hr) and resistance to winter dieases.

Score card for frost damage

Rosemonde Devaux, Isabelle Lejeune-Hénaut

At INRA, frost damage is scored within block trials constituted with individual plots of six rows generally repeated three times. It is recommended to use only the four central rows for scoring to eliminate the border effect (due to the wheels of the drill).

Damage should be scored about 10 days after a period of frost to better identify frost burn. The score should be attributed to the plot as a whole and based on the appearance of the aerial parts of the plants according to the scale described below.

Field scoring scale

Score	Damage to the plant
0	No damage
1	Leaf border affected (resembles burnt paper)
2	Majority of leaf surface affected
3	Partial damage to stem (upper third of plant)
4	Serious damage to stem (upper three-quarters of plant)
5	Damage to whole plant

Intermediate scores should be used to record irregular damage within a plot. This can be done by averaging several scores attributed to different areas of the plot. For example, for a plot in which surface-equivalent areas were given a score of 1 and 2, the average score would be 1.5.

Additional information is very useful for analyzing results:

— a record of plant emergence is indispensable as some genotypes may otherwise be wrongly scored due to late emergence;

— four regular samplings should be made on control varieties during the course of vegetation. The following information should be noted:

• stage of development of the plant: number of nodes, branches, height in cm;

• condition of the stem, leaflets and, if possible, of the roots; presence of burns or necrosis;

- condition of the apex (present, absent, reduced);
- developmental stage of the apex (vegetative or reproductive) observed with magnifier glasses;
- presence of diseases associated with frost (in certain locations). *Ascochyta* blight should be scored using the Tivoli scale (1994) (see fig. 2.22), and the presence of bacterial blight should be recorded.

If temperatures are below −10°C and the soil is waterlogged, it is important to record roots uncovered by frost heaving. At the end of winter, around mid-April, regrowth ability and the rosette should be scored (0 to 3 scale: 0 = upright stem with no branching, 3 = rosette habit with numerous primary and secondary branches).

7

Biotic stresses

Impact of aschochyta blight on spring PEA functioning yield[1]

Bernard Tivoli, Christophe Le May, Alexandra Schoeny, Marie-Hélène Jeuffroy, Bertrand Ney

Yield losses caused by diseases are generally difficult to predict with accuracy. In order to characterize the quantitative and qualitative impacts of Ascochyta blight on pea crop, studies involving collaborations between pathologists, ecophysiologists and agronomists aim to determine the incidence of ascochyta blight (i) on the synthesis and transfer into the seeds of carbon and nitrogen compounds at the individual pea plant level and (ii) on yield components at the plant population level. The two main questions are: first, how is plant function disturbed by the disease? Second, what level of disease is associated with yield loss? These studies help to improve our understanding of the cause of the damage, and should reveal the effects of essential factors such as:

- The period of infection, by determining the incidence of the disease for various stages of canopy development;
- The site of infection, by determining the incidence of the disease when it develops on leaves, stems and/or fruiting organs, and according to the growth stage of the affected plants;
- The intensity of the infection, by quantifying relationships between the various levels of disease and yield loss.

---

1. This article is based on the article: Tivoli B, Ney B., Jeuffroy M-H., 1999. L'anthracnose du pois protéagineux. Mieux évaluer sa nuisibilité pour mieux raisonner la protection de la culture. Phytoma La défense des Végétaux, 512, 16–20.

We aim here to summarise the main results obtained to date concerning studies of the damage caused by this disease and to show how these results can be exploited in broader plant protection approaches.

The pathosystem

The three elements of the pathosystem pea/*Mycosphaerella pinodes* (the plant, the pathogen and the environment) must be taken into account to understand the impact of the disease on plant functioning and yield elaboration.

• Plant and plant canopy

Diseases evolve within plant populations, characterised by particular patterns of growth and development. The growth and development models established by agronomists apply to the functioning of such plant populations in optimal conditions, i.e. in the absence of abiotic or biotic constraints (such as diseases).

The functioning of the pea canopy has been described in detail in the previous chapters (see part I). Only a few key elements will be repeated here. Grain yield depends on the carbon (biomass production) and nitrogen (symbiotic fixation, remobilisation) metabolic pathways in the plant. The vegetative development of the plant leads to the establishment of organs subsequently involved in plant growth (see chapter on vegetative development in part I). The amount of biomass produced depends on the total amount of radiation reaching the crop, the proportion of this radiation intercepted by the canopy and its conversion into biomass (Ney, 1994; see chapter on carbon acquisition at the crop level in pea). Nitrogen is generally accumulated through the symbiotic fixation of nitrogen gas from the atmosphere, with most nitrogen accumulated before the start of the seed-filling period (see part on C and N fluxes within the plant). The accumulated nitrogen is stored in the vegetative organs and is then remobilised and transported to the seed during seed filling (see p. 116 and p. 119 to 121). The number of seeds produced by the plant depends directly on the growth rate between the beginning of flowering and the end of the final stage in seed abortion (see p. 105). Mean seed weight also depends partly on biomass accumulation during the seed filling and on nitrogen resource remobilisation periods (see p. 116–117).

• The pathogen and the disease

Ascochyta blight, which is generally most severe after flowering, is described as an "end of vegetative cycle" disease on spring pea crops. It is caused by a necrotrophic fungus with two forms: an asexual form producing pycnidia (*Ascochyta pinodes*) and a sexual form producing perithecia (*M. pinodes*).

The first symptoms consist of flecks in leaves and pods, and stripes on the stems. The flecks then take on the appearance of angular necrotic lesions, generally with distinct boundaries. These necrosis may extend over the entire surface of the organ following the weakening of plant defences. Stem striping may also progress towards continuous lesions, girding the stem on various heights.

It has been shown under controlled conditions that the onset and development of the disease are tightly linked to temperature: at 20°C, infection is rapid, as the fungal spores germinate within six hours and the first symptoms appear in 24 to 48 hours (Roger et al., 1999).

The pathogen spreads by producing pycniospores (dissemination by splashing, in small drops of rainwater, over short distances) and ascospores (air-borne dissemination over longer distances). Pycnidiospores are produced by asexual reproduction and serve to establish the fungus on the plant. Ascospores are produced by sexual reproduction and are responsible for the rapid increase in disease levels after flowering in pea (Bretag, 1991; Hare and Walker, 1944; Roger and Tivoli, 1996). A simplified disease cycle is presented at figure 2.20.

Observations over a number of years have demonstrated that the intensity of the disease on individual plants varies with height. Disease scores for different leaves, internodes and pods show that the disease is most severe at the base of plant, moderately severe at intermediate levels and least severe or non-existent at the top of the plant (fig. 2.21).

The 0–5 disease scale proposed by Tivoli (1994), takes into account disease progression on the various organs of the plant (fig. 2.22).

- Climate

The effects of climatic environment on climatic aspects, linked to disease development, the formation of fruiting bodies and parasite dissemination have been extensively discussed in the articles by Roger and Tivoli (1996) and Roger et al. (1998). We will recall here simply that the severity of the disease increases with humidity and is greatest at temperatures between 15 and 25 °C.

Overall effects of ascochyta blight on yield and yield components

A field experiment involving "healthy" plots, protected by fungicide applications and with plots artificially contaminated with the fungus, investigated the effect of the disease on yield and yield components (Tivoli et al., 1996). The disease does not affect all yield components in the same way, as shown by Béasse (1998) and Le May (2001). Stem number and height, the number of fruiting nodes and the number of pods are not affected.

Secondary contaminations

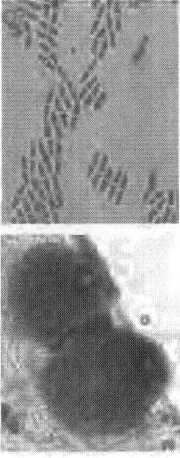

Pycniospores

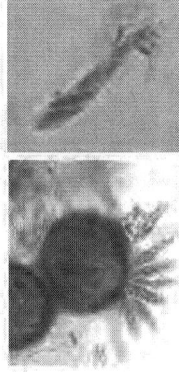

Pycnids

Asque et ascospores

Périthèces

Primary contaminations

- Soil
- Residues of non buried crops
- Pea regrowth
- Wild leguminous plants

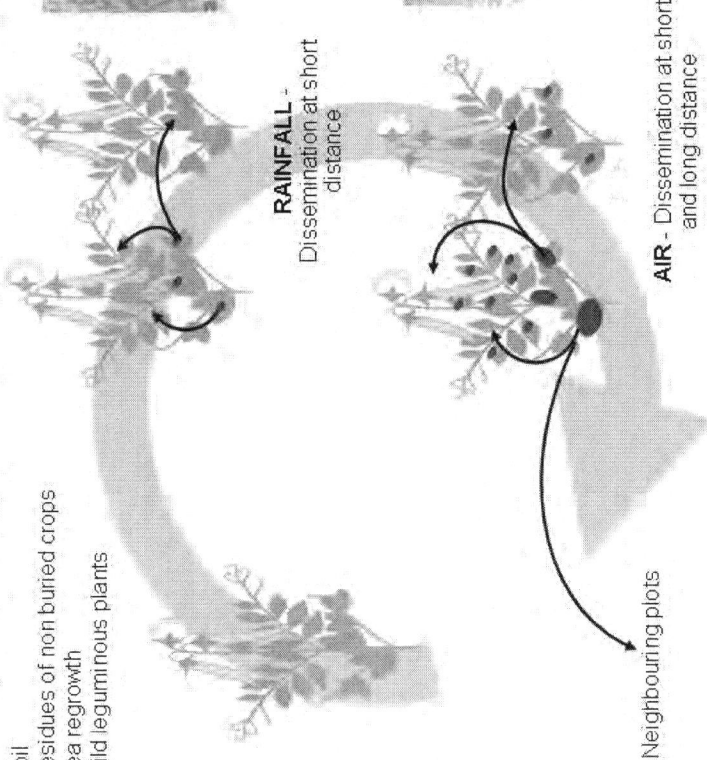

RAINFALL - Dissemination at short distance

AIR - Dissemination at short and long distance

Neighbouring plots

Figure 2.20. Cycle of Micosphaerella pinodes.

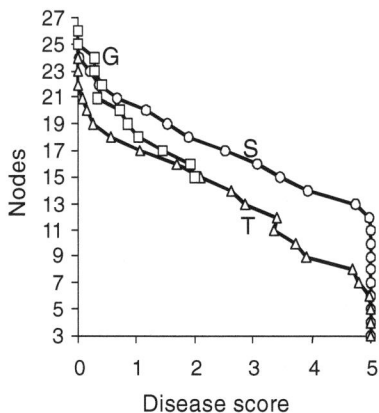

Figure 2.21. Disease score profile on stipules (S), internodes (I) and pods (P), at the various growth stages of the plant.

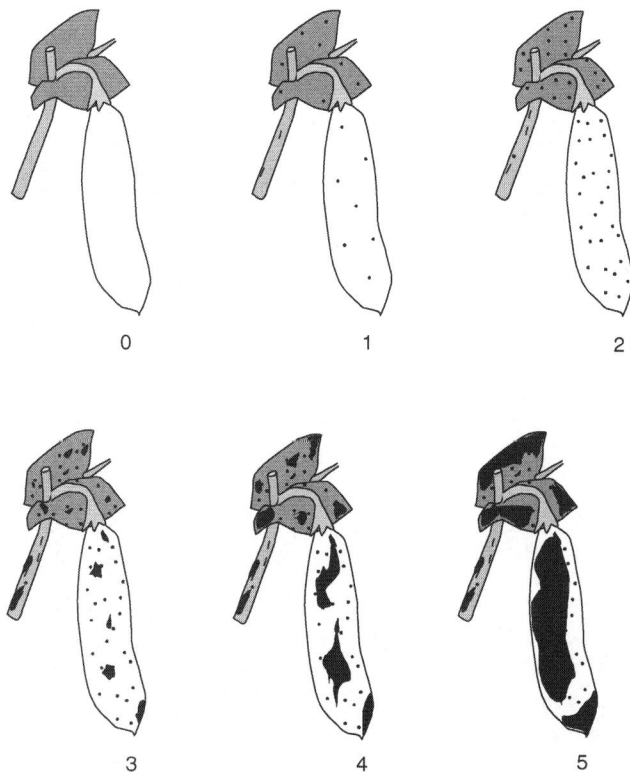

Figure 2.22. Scoring scale for ascochyta blight due to M. pinodes (Tivoli, 1994).

In contrast, the number of seeds per stem, mean seed weight and harvest index, are all significantly decreased, by 21.5 %, 15.5 % and 17.5 %, respectively for a typical plant density of 80 to 90 plants per m². Biomass production is also 20 to 30 % lower for diseased than for healthy plants. The final yield losses observed in diseased plots may be as high as 30 to 40 % of the yield obtained from plots with total fungicide protection.

Effects of the pathogen on host physiological processes

The development of symptoms is associated with changes in the physiology of the plant, including in particular photosynthetic activity and carbon and nitrogen remobilisation activities.

• Photosynthetic activity

Photosynthesis is the main process determining the overall plant metabolism. This process is responsible not only for capturing energy from light, but also for synthesizing carbon compounds and other molecules essential to the plant. Garry *et al.* (1998b) showed that during their interaction the necrotrophic fungus, *M. pinodes*, markedly reduced photosynthesis in the leaves of a pea plant and that the magnitude of this decrease increased with the level of infestation (table 2.3).

Table 2.3. Impact of ascochyta blight (disease scale 0–5) on the photosynthetic activity of pea stipules (expressed as a percentage of the value for a healthy control), for leaf discs and for the whole plants (from Garry *et al.*, 1998b).

Disease score	0	1	2	3	4	5
Leaf disc % reduction %	0	7–18	13–35	53–56	92–98	100
necrotic area	0	0	ε	8	17	29
Whole plant % reduction	0	0	25	40	100	100

The causes of this decrease in activity remain unclear, although two explanations have been suggested: (1) photosynthesis is disturbed by the spread of a toxin (ascochytin) synthesized by the fungus in cells adjacent to the necrotic area; (2) pathogen infection induces plant defence mechanisms including the production of phytoalexins, which may have deleterious effects at high concentration (Heath and Wood, 1971; Kué, 1995; Lepoivre, 1982; Van den Heuvel and Grootvel, 1980).

Thus, the effects of the disease on photosynthetic activity extend far beyond the simple necrotic area. Several studies (Bassanezi *et al.*, 2001; Bastiaans, 1993; Madeira and Clark, 1995; Scholes, 1992; Wright *et al.*, 1995) have tried to link organ function (photosynthesis in a leaf, for example), with the extent to which the organ concerned is attacked. For rice blast (caused by *Pyricularia oryzae*), Bastiaans (1991) demonstrated that the decrease in photosynthesis of a diseased leaf (P_x/P_0, where P_x and P_0 are the levels of photosynthesis of the same portion of a healthy and diseased leaf, respectively) can be related to the percentage of necrotic leaf area (x): $P_x/P_0 = (1-x)^\beta$.

For rice blast, the coefficient β is much higher than 1 (about 3.5), suggesting that this disease has a much larger effect than suggested by the proportion of diseased tissue. The disease therefore seems to have a profound effect on photosynthesis phenomenon beyond the observed necrosis. This led Bastiaans (1991) to introduce the concept of "virtual lesion." The relationship between the virtual and real necroses increases in strength with increasing coefficient β.

This formalism allowed to classify diseases according to the decrease in photosynthetic activity they are likely to cause (Bastiaans, 1991). Thus, on winter wheat, brown rust (*Puccinia recondita*) has an effect almost proportional to the residual green area of the necrotic leaf area ($\beta=1.26$), whereas powdery mildew (*Erisyphe graminis*) has a much greater effect than suggested by the necrotic area ($\beta=8.74$).

Garry *et al.* (1998b) showed that the effect of *M. pinodes* on photosynthetic activity is twelve times greater than the necrotic area ($\beta=12.1$). The disease affects the interception of radiation by causing necrosis and inducing the premature senescence of leaves. It also affects the conversion of this radiation into dry matter, by reducing net photosynthesis and, undoubtedly, by diverting some of the assimilates produced by the plant to the pathogen.

- ## The remobilisation of carbon and nitrogen

An alternative explanation for the decrease in photosynthesis following the development of the pathogen on the leaf involves remobilisation of the products of light energy transformation (sugars and proteins). Photosynthesis is subject to a system of self-regulation. Imbalance in the system can rapidly lead to the inhibition of photosynthetic activity (Scholes, 1994). Experiments involving plants cultivated in greenhouses and presenting various levels of disease showed that the disease affects not only photosynthetic activity, but also the remobilisation of nitrogen from the vegetative organs to the seeds (Garry *et al.*, 1996). Indeed, the concentration of nitrogen compounds in diseased stipules and pods is higher than that in healthy organs, and this difference is particularly marked if the disease is severe (table 2.4): nitrogen seems to be "retained" in the diseased organs and its transfer to the seeds is blocked.

Table 2.4: Impact of ascochyta blight on the quantity of nitrogen in the form of protein at harvest (% of healthy control), in the stipules of plants inoculated by spraying with spores at three concentrations.

Concentration of spores	0	10^4	10^5	10^6
Disease score	0	2.9	3.5	4.5
Protein nitrogen	100	160	270	380

This disturbance of both photosynthetic activity and nitrogen remobilisation results in a decrease in the number of seeds and in individual seed dry weight. In the seeds, carbon metabolism seems to be affected by the disease to an even greater extent than nitrogen metabolism (Garry *et al.*, 1998; Béasse *et al.*, 1999).

The effect of plant growth stage and disease location on damage to the plant

The impact of the disease on yield is not fixed and depends on growth stage at the onset of disease and the location of the disease on the plant.

• Plant growth stage

The indeterminate growth habit of pea plants results in the simultaneous presence on these plants of organs of different types and different ages. This property is of particular importance because behaviour with respect to the disease may differ according to the age of the organ concerned. Two stages seem to be particularly important for yield development: the beginning of flowering (BF) and the final stage in seed abortion (FSSA) (Garry *et al.*, 1996; table 2.5).

Table 2.5. Impact of ascochyta blight on seed number per node, individual seed weight and decrease in total seed weight per node (as a % of the healthy control), in plants inoculated by spraying with a suspension of spores (10^6sp.ml^{-1}), at flowering or at the final stage in seed abortion (FSSA) (from Garry *et al.*, 1998a).

	Seed number		Mean seed weight		% decrease
	Flowering	FSSA	Flowering	FSSA	
Healthy	3.1a	5.4a	0.358a	0.345a	66
Diseased	1.1b	5.8a	0.355a	0.305b	4.8

• a, b : values followed by a same letter are not significantly different at P=0.05.

- The decrease in total seed weight per plant is particularly large if the plant is inoculated early; similarly, the yield loss for an entire plant inoculated at flowering of the second node is 68% whereas that for a plant inoculated at beginning of seed filling (BSF) of the first node is 40 %.
- If the plant is infected with M. pinodes at flowering, the observed decrease in yield results from a decrease in seed number.
- If the plant is infected after FSSA, the number of seeds is not affected, and the observed yield loss results from a decrease in individual seed weight.

• Location of the disease on the plant

Ascochyta blight develops on the stipules, stems and pods. The relative impacts of disease on different organs on plant functionning and yield loss remain unclear.

However, Béasse (1998) showed that ascochyta blight occurring on leaves at the base of the plants after canopy closure did not cause yield loss. Indeed, these leaves, which are already old and receive little light by this stage, are already beginning to senesce and play no further role in plant growth. However, these findings should not be interpreted as indicating that these symptoms have no epidemiological effect.

It has been shown that symptoms limited to the stem only reduce yield if the infection is very severe and if the disease extends a long way up the stem (at least 40 % of the stem displaying severe necrosis). Such symptoms are generally observed at the time of physiological maturity and may, in extreme cases, be highly damaging. They accelerate lodging by weakening the stem.

Finally, Béasse et al. (1999) showed that disease on the pods may cause yield loss, but that this loss accounts for no more than 10 to 15 % of total yield loss. The remaining loss is due to a decrease in growth due to the decrease in photosynthesis and, in some cases, poor nitrogen remobilisation.

The use of this knowledge for canopy functioning modelling

• Construction of a model of diseased canopy functioning

The aim is to introduce the disease and its effects into models of diseased plant canopy such that damage can be predicted as a function of the type of epidemiological development. This type of model (Béasse, 2000), based on the work of Monteith (1972), has been used to calculate on the reduction in biomass production in a diseased canopy by:

- Estimating the contribution of each node to growth (Ci) as a function of its contribution to the interception of radiation.
- Calculating the decrease in the contribution of each node due to disease, using the observed linear relationship $P_x/P_0 = ax + b$, where a and b are constants and x is the disease score for the node in question.
- Adding together these individual contributions to determine the total decrease in biomass production due to disease.

The results obtained show that the damage caused to plant populations by a pathogen progressing from the base to the apex of the plant depends on the capacity of the plant to preserve its "active" nodes for photosynthesis. Crops with high levels of infection may show little yield loss if the plant manages to maintain the healthy leaf area (or root mass for soil diseases) required for full light interception. The effect the disease on the plant photosynthetic activity depends on the disease score profile observed at a given time (fig. 2.23). In 1996, the upper nodes of the plant producing a large leaf area, make a major contribution to canopy function. The disease score profile was also higher in 1996 than in 1995. The conjunction of disease and leaf area profiles led to 30 % and 50 % decreases in photosynthetic activity in 1995 and 1996, respectively. These changes resulted in a decrease in growth rate only in 1996 (about 20 %) and in a decrease in yield of 15 %.

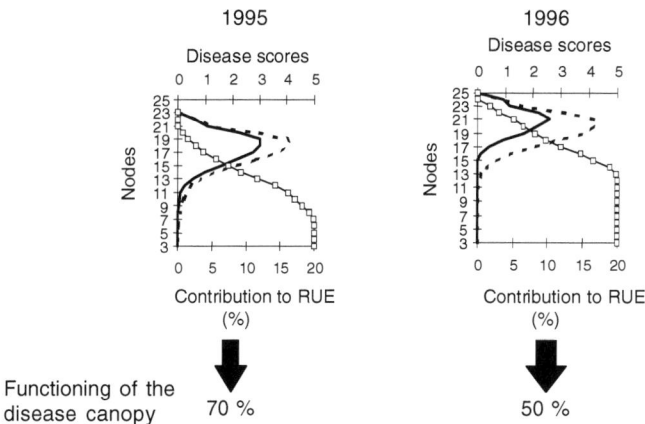

Figure 2.23. Impact of ascochyta blight on the photosynthetic activity of each node of the plant (determined at the end of the final stage in seed abortion) in 1995 and 1996 at Le Rheu. The photosynthetic activity of the diseased (full line) and healthy (dotted line) nodes is shown. Disease scores are plotted as open squares.

• Effect of cultivar on damage

Each year, new cultivars were grown in France and elsewhere in Europe. These genotypes differ in terms of their potential yield, stem habit or flowering time. However, differences in canopy architecture (branching ability, stem height, standing ability) may have an impact on the development of the disease and its effects on photosynthesis and on the resulting effects on yield (Le May, 2001). Six cultivars susceptible to ascochyta blight (Aladin, Athos, Baccara, Bridge, Obélisque and Solara) showed identical disease profiles at individual plant level but the progression of the disease differs between cultivars at canopy level: the disease progressed more rapidly on Aladin crops than on crops of the other cultivars. Stem density, stem height, node number and the distribution of leaf area as a function of node contribute to the establishment of a particular canopy architecture that may influence the microclimate (modifying temperature and leaf wetness duration), thereby also affecting the development and dispersal of the disease (through splashing and wind turbulences).

The disease had similar effects on photosynthetic activity in all six genotypes, but differences in biomass production were observed between the cultivars due to differences in canopy structure resulting from, in particular, the distribution of leaf area as a function of node. The gradual progress of the disease up the plant with time and the distribution of the photosynthetically active area resulted in differences in the field behaviour of the six cultivars under the same disease conditions.

Conclusion: combine the knowledge of the impact of ascochyta blight on yield in the integrated disease management strategies

Plant pathologists generally adopt one of two distinct approaches to studying aerial diseases appearing in the field and developing in the plant canopy: epidemiology in the strict sense of the word (the most frequently used approach) and approaches focusing on damage.

For ascochyta blight on pea, a combination of these two approaches has been used: the integration of epidemiological data (providing information about the risk factors for the development of an epidemic) into a model simulating the effects of the disease on the growth and development of pea plants. This approach takes account of disease severity, plant growth stage and the risks of epidemic development, and should make it possible to use treatments more effectively (by improving treatment timing) and consequently reduce the number of fungicides applications. Furthermore, these observations should lead to a better typing of cultivars in terms of the role of their architecture in the development of disease epidemics and possible tolerance (particularly with respect to the ability to maintain normal levels of photosynthesis following infection). Several recent studies have shown that plant and canopy architecture have an important effect on ascochyte blight epidemics (le May *et al.*, 2009a, 2009b; Schoeny *et al.*, 2008). Finally, the studies cited here were carried out with only one of the three agents causing ascochyta blight, *M. pinodes*. As pointed out by Allard *et al.* (1993), several species are likely to coexist with this pathogen (*Phoma medicaginis* var *pinodella*, *A. pisi*). The real questions to be addressed therefore concern the way in which this complex of species develops in the canopy (in time and space) and the damage such a complex is likely to cause. This approach, which has rarely been used to date, should make it possible to adapt cropping techniques to the farming characteristics of particular regions (varieties used, sowing densities etc.).

Sitona

Yves Crozat, Thierry Doré

The pea weevil (*Sitona lineatus* L.), a coleopteran of the Curculinidae family, has long been wrongly considered to be a minor pest of pea. In France, it is now recognised that damage caused by this parasite can severely affect yields. Levels of loss seem to vary considerably between situations and maximum losses reported are about 1 to 1.5 t/ha (Taupin *et al.*, 1994). There are two main difficulties associated with the analysis of *Sitona* effects on pea yield in field conditions. Firstly, though leaf damage caused by adults during crop growth is relatively easy to quantify, the same cannot be said for damage to underground parts caused by larvae emerging after adults have laid eggs. Secondly, assessing the effects of an altered root system on yield development and quality is a complex issue.

Life cycle, infestations and the assessment of their damage, and the influence of cropping systems

In order to assess infestations and the damage they cause, a good knowledge of the pest's life cycle is essential. Only one generation of weevils is produced each year, but adults (with a life expectancy of around 10 months) that appear on a pea crop in one year, die on the following year's crop. Crops are therefore infested by adults that have over-wintered in sheltered places (woods, ditches etc.), which they fly away from as soon as temperatures reach a threshold of between 10 and 15 °C. These adults seek out pea plants (or other leguminous plants) upon which to feed and reproduce, using their olfactory sense (Landon *et al.*, 1997). Once settled, the first adults attract mates using a pheromone-based aggregation system (Blight *et al.*, 1984). It is at this point that adults feed on stipules and leaves, causing damage to foliage. Afterwards, these adults lay their eggs in the soil and die. After hatching, the larvae from these eggs feed on nodules or even roots. Following a short nymph stage, the adults emerging at the surface of the soil towards the end of the pea growth cycle leave to search for food before finding shelter for the winter. Survivors start a new cycle on the following year's pea crops.

Adult population sizes can be assessed indirectly from damage to foliage. In this way Cantot (1986) developed a scoring system based on the number of notches per pea stipule; this system was simplified by Doré and Meynard (1995) in order to enable rapid assessment on a large number of plots. Larval population sizes, rather like damage to root nodules, are much more difficult to estimate. At the very least, root systems should be extracted to obtain an estimate, and ideally nodules should be weighed. However, Cantot (1986) found a relationship between adult population sizes and the larvae of the following generation under semi-controlled conditions, and Doré and Meynard (1995) showed an acceptable correlation between foliar damage and nodule damage on untreated plots.

Aside from insecticide treatments, the influence of the cropping system on the intensity of attacks remains relatively unclear; the effects of soil tillage techniques in particular give cause for debate. Nonetheless, it is known that early sowing dates increase the probability of intense attacks (Doré and Meynard, 1995; Steene and Vulsteke, 1999): indeed adults cease to move around after finding a first feeding site, generally preferring to remain on the plot within which they first landed. The first crops sown therefore attract nearly all the adults. However, when there are several flights of weevils, as occurs quite frequently, the later crops may also be affected. Neither the size of the area devoted to peas (Hamon *et al.*, 1987), nor the variety (Steene and Vulsteke, 1999) seems to influence the intensity of attacks, which vary from year to year, with winter and spring weather conditions certainly playing an important role in this variability. Finally, there may be wide variability in the intensity of attack, even within a plot (Doré and Meynard, 1995; Schotzko and Quisenberry, 1999), making monitoring more difficult.

Effects on nitrogen use and yield development

Landon *et al.* (1995) estimated that an adult could consume the equivalent of a notch per day, resulting in the loss of approximately 6 mm^2 of leaf area. At the crop scale, this foliar loss might seem insignificant given the rapid growth of the photosynthetic surfaces during the vegetative stage. Indeed, under controlled conditions, by introducing only adult males (13/plant at the one leaf stage), Ravn and Jensen (1992) showed that leaf bites did not significantly affect pea growth; however Livy Williams *et al.* (1995) observed yield losses of up to 10 % when infestations were intense during crop development.

Most of the available work considers the damage caused to the root system by larvae, as this is much more severe than the foliar damage and leads in particular to the destruction of root nodules. Larval infestations coincide with the formation of the N-fixing system, which attains its maximum size (in terms of biomass and number of nodules) around the beginning of flowering (Tricot-Pellerin *et al.*, 1994). As far as we know, there has been little direct quantification of the effects of nodule damage on nitrogen fixation. Several studies have demonstrated a good relationship between the biomass of active nodules and nitrogen fixing activity and/or nitrogen status of pea crops (Crozat *et al.*, 1991; Doré and Meynard, 1995). It is therefore not surprising to note that instantaneous nitrogen fixation activity (a measure of nitrogenase activity based on the acetylene reduction method) decreases with increasing proportion of nodules damaged by weevil larvae (fig. 2.24).

According to figure 1, an attack affecting 22 % of nodules can lead to a 50 % decrease in nitrogen fixation activity. At the crop scale and in the field, Corre-Hellou and Crozat (2004) found greater losses by measuring nitrogen fixed using a comprehensive method based on natural abundance of ^{15}N. Thus a survey of farm fields showed that for plots with a mean score for foliar damage of 3 (according to the Cantot (1986) scale) at the 3–5 leaves stage, the nitrogen fixation rate at harvest was never higher than 25 %. When the foliar damage score is less than or equal to 1, fixation rate is close to 80 %. The Nitrogen Nutrition Index (NNI), measured at the beginning of flowering, decreases with increasing foliar damage (fig. 2.25). However, within each weevil leaf damage score there is considerable variability in NNI.

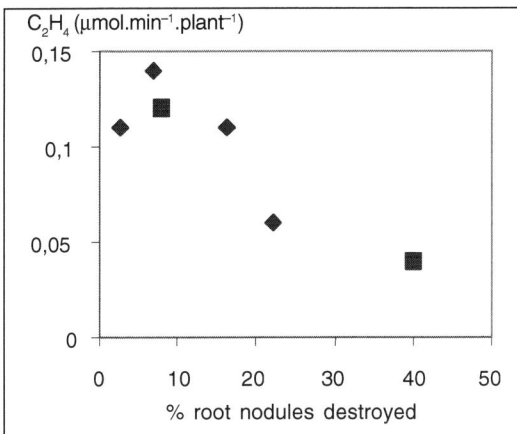

Figure 2.24. Influence of root nodule destruction rate by *Sitona* larvae on instantaneous nitrogen fixation activity measured by acetylene reduction (after Ravn and Jensen, 1992) ◆ pot experiment ■ field experiment with or without seed treatment.

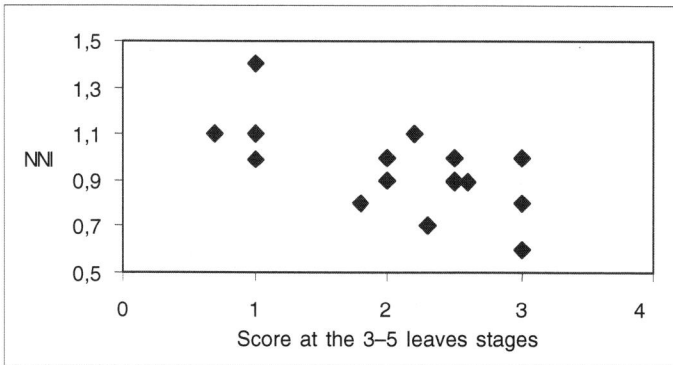

Figure 2.25. Foliar weevil damage scores observed at the 3-5 leaf stage (Cantot scoring system) and Nitrogen Nutrition Index (NNI) measured at the beginning of flowering in a network of organic pea plots in Pays de Loire, France (1999–2000). Only those plots with low weed infestation (< 25 g MS/m² at the beginning of flowering) and without water stress are presented (Corre-Hellou and Crozat, 2004).

A maximum score does not always equate with nitrogen deficiency (INN< 1). The NNI takes into account both nitrogen fixed and mineral nitrogen in the soil used up until flowering. In situations where the soil provides large amounts of nitrogen, this indicator of crop nitrogen status may intervene too early to detect alterations to the fixation system caused by weevils.

Doré and Meynard (1995) studied a network of farmers' fields over several years and also observed a decrease in NNI at the beginning of flowering related to *Sitona* damage on root nodules. Even though both studies cited above (Corre-Hellou and Crozat, 2004; Doré and Meynard (1995) were conducted in very different soil and weather conditions, with different cropping systems, in both the authors noticed a positive relationship between the NNI and the number of seeds per m² caused by an influence of nitrogen nutrition on the ramification ability of plants and/or on the number of seeds per stem (Doré and Meynard, 1995; Corre *et al.*, 2001). In addition to these effects of weevil damage on the number of seeds, there is also an effect on seed protein quality. As Bourneville *et al.* (1990) have shown over several years, there is a good relationship between protein concentration of seeds and adult weevil attacks.

These results show that understanding the effects of Sitona on yield depends on an accurate evaluation of the influence of the damage on the nitrogen status of the crop.

Used in conjunction with other tools which can simulate the effects of nitrogen nutrition on yield development and on pea protein quality (see preceding chapters), this type of evaluation should facilitate the development of a functional model for pea which integrates the effects of this parasite.

Weeds

Yves Crozat

A large number of references are now available concerning the management of weeds in pea crops (Arvalis—Institut du végétal 2003). Removal of hardy grasses and annuals is not difficult in pea and Faba bean crops as specific herbicides exist. Otherwise, if persistent grassy weeds are present, sowing a legume crop in spring can provide a non-chemical solution for eliminating weeds that have colonised during the winter. Controlling certain dicotyledons is sometimes more challenging, particularly after crop emergence, and especially where Faba bean crops are concerned, as few products are available.

Therefore in practice, in conventional farming, while yield losses due to weed infestation (direct damage) are practically negligible, the risks of weed contamination at harvest and/or of secondary damage (increased seed bank of certain weeds) are substantial.

Conversely, in systems that do not use herbicides (organic agriculture), or in situations where weed control fails, weed infestations are one of the major factors limiting pea yields (Corre *et al.*, 2001). During several years of monitoring of spring pea plots cultivated organically, the authors estimated that in nearly 20 % of situations, yield losses could reach 50 % (of a potential yield estimate). In such situations, weeds mainly affect the number of seeds per stem as a result of a strong decrease in nitrogen nutrition. These effects are visible at the beginning of flowering, when Nitrogen Nutrition Index (NNI; p. 61 and p. 64) values may drop below 0.6. Since weeds are more competitive than pea at extracting mineral nitrogen from the soil, pea nitrogen nutrition depends almost entirely on symbiotic fixation. Despite this, due to further competition with weeds for light interception, fixation potential is limited and therefore insufficient to meet the peas' needs (Corre-Hellou and Crozat, 2004).

Pea is often cited as a poorly competitive species compared with weeds (Wall *et al.*, 1991; Townley-Smith and Wright, 1994). At least two factors might explain this weak competitiveness: (i) slow and shallow root development, putting pea at a disadvantage with regard to nutrient access compared with weeds (Hauggaard-Nielsen *et al.*, 2001; Hellou and Crozat, 2004), (ii) relatively slow leaf growth, especially in the case of winter pea.

The percentage of light intercepted by a pea crop is a useful variable which can indicate its competitive ability (cf. part "Carbon acquisition at the crop level in pea"). The competitiveness of a pea crop can be evaluated by the ratio of "weed biomass associated with peas/weed biomass without peas". For a given cultivar, this ratio increases with increasing ability of the crop to intercept sunlight during the vegetative stage (fig. 2.26).

% decrease in weed biomass at harvest

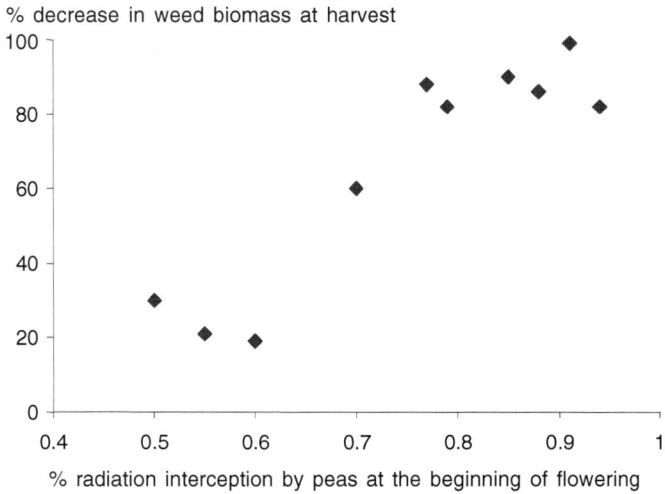

% radiation interception by peas at the beginning of flowering

Figure 2.26. Proportional decrease in weeds (%) measured at harvest (100 – (biomass of unweeded quadrats with pea crop/biomass of weeds in adjacent quadrats without pea crop)) as a function of percentage radiation interception at the beginning of pea flowering. Pays de la Loire-ESA network-2001.

Several factors give rise to this competitiveness:

- number of seeds sown;
- percentage and speed of crop emergence, both of which in turn depend on seed quality, soil preparation and weather;
- competition for light between the crop and the weeds, as well as for water and nitrogen, may regulate the rate of leaf expansion.

In addition, the competitiveness of pea towards weeds can vary significantly between cultivars (Wall and Townley-Smith, 1996; Crozat *et al.*, 2000). The most competitive cultivars tend to produce nodes very rapidly, to have a high leaf area index leading to rapid canopy closure, to grow tall and to have strong stems. An example of such a cultivar would be Nitouche, as compared with the less competitive Baccara (Crozat *et al.*, 2000).

PART III

Integration of knowledge into a global model and examples of application

8

A model which integrates knowledge on pea crop physiology and agronomic diagnosis

Véronique Biarnès, Jérémie Lecoeur, Marie-Hélène Jeuffroy, Aurélie Vocanson, Bertrand Ney

Why develop a pea crop model?

Crop models bring together the knowledge obtained on a species for a specific use, and therefore are useful for identifying gaps in our knowledge of the physiology of the species concerned. They may be used to simulate the behaviour of crops grown without any limiting factors in order to make an agronomic diagnosis at the plot scale. In this case, the objective is to answer questions such as: what was the yield potential of the plot allowed by the climate? Did some factors limit the yield? Which ones? When did they affect the crop?

Crop models may give some valuable information to quantify the impact of limiting factors on the crop yield, by coupling the use of crop models with measurements on plants, often expensive. Thus, the comparison of simulated meteorological indicators with yield measured on a crop allows rapid progress to be made in the identification of critical periods for the crop, and of the limiting factors (Doré *et al.*, 1998).

Earlier in this book, knowledge is presented on the physiology of the pea plant on different scales, from parts of organs to the whole canopy (see parts I and II). These results provide quantitative information on numerous processes which interact with the environment. However, these analytical approaches do not allow the interactions between all the studied processes which determine the crop yield to be predicted.

To account for the consequences of multiple interactions between different components of the environment and states of the crop, it is necessary to bring together this information in a crop growth model in order to estimate the crop yield, nitrogen balance, seed quality, water consumption, etc.

The number of environmental factors which can be taken into account depends on the state of the available knowledge. The information collected so far enables us to propose a crop growth model for spring field pea which includes the following factors: temperature, solar radiation, water and nitrogen availability. The state of our knowledge does not yet allow us to include biotic constraints. For some modules, specificities of the varieties may be taken into account using a parameter set collected from networks of variety trials.

The objective of this crop model, which has been developed for peas, is twofold. On the one hand it provides scientists working on field peas with a tool which integrates existing knowledge, and on the other it must also be sufficiently practical to respond to the needs of agronomic diagnosis. This may represent a difficult gamble because it lies mid-way between a cognitive work and an engineering work (Passioura, 1996). We feel sure that the continuous dialogue in these phases of conception of the model between scientists and those responsible for technical support will ensure that appropriate choices are made to respond to this double objective (cf. p. 19 and 20 (beginning of part I)).

General structure of the crop growth model and minimum data required

Crop growth models are always made up of several modules. These modules communicate with one another by means of state variables which characterize the plant or the environment. The main state variables are crop development variables, leaf area index (LAI) of the crop, root depth, aerial biomass, amount of nitrogen accumulated by the canopy and the amount of soil water available for the crop.

The work consisted firstly elaborating a model for non-limiting conditions, then adding modules taking account of abiotic constraints. The modules created to simulate the physiology of a pea crop, with only temperature and solar radiation as limiting factors, are:

— a development model;
— a module for setting up aerial captors;
— a module of biomass production;
— a module of yield component elaboration.

To take into account availability of water, we must add to these modules:

— a module for water balance;
— a module for the evolution of underground captors;
— a module of distribution of nitrogen in the plant.

These different modules are inter-connected in order to account for interactions between different processes depending on their response to the environment. The complexity of the modules is variable, ranging from a single equation to a complex algorithm. This complexity depends on the degree of stability of the simulated processes or on the number of variables which affect them.

To estimate the state variables of the model, it is necessary to provide information which describes the simulated system and the input data needed to initialize the model. To describe the system, the minimal information required is: the location (latitude), the sowing date and a simple soil description including at least the texture and depth of soil accessible by the roots. The variety can also be described with more or less detail. Information on the developmental calendar for the variety are generally supplied in the model.

The input data needed to initialize the model are essentially the moisture status of the soil. Furthermore, daily meteorological data are necessary because the model runs on a daily time step; they include mean air temperature, solar radiation, reference evapotranspiration and rainfall. It may be necessary to add information about cultural techniques such as water supply.

Description of modules

The model below corresponds firstly to modules describing the growth of a spring pea crop in non-limiting conditions, designed by Ney (1994) on the basis of work presented in the first part of this book and secondly, to modules that simulate the effect of abiotic stresses (drought and high temperature), based on results of the work of both Lecoeur and Sinclair (1996) and Guilioni (1997), shown in figures 2.1 and 2.8. This model became a program named Afila. Vocanson (2006) has adapted it for autumn sowings.

Module of phenology

The phenology module can simulate the five following stages: emergence, beginning of flowering (BF), beginning of seed filling (BSF), final stage in seed abortion (FSSA) and physiological maturity (PM).

• Date of emergence

The date of emergence is simulated by adding 150 degree-days (base temperature: 0°C) to the sowing date. This is valid for sowings from January to April. Outside this period, particularly for autumn and winter sowings, this figure does not apply because it under-estimates the duration of the "sowing-emergence" period.

• Date of beginning of flowering

The date of the beginning of flowering is calculated from the model of Summerfield and Roberts (1988) (cf. p. 41 (à voir)), parameterized by Roche (1998). It depends on mean temperature and mean photoperiod during the period "emergence-beginning of flowering":

$$1/f = aT + bP + c$$

where f = number of days from emergence to BF; a = 0,0013347; b = 0,0015626, c = −0,020174

With this equation, the prediction of the date of beginning of flowering is correct for sowings made from January to May. To simulate the date of beginning of flowering for autumn sowings or other types of genotype, particularly photoperiod-sensitive ones, it is necessary to modify this model.

• Date of beginning of seed filling

The date of beginning of seed filling occurs 250 degree-days after the beginning of flowering (Ney and Turc, 1994).

This parameter exhibits weak genotypic variability (Dumoulin *et al.*, 1994). Indeed, these authors estimated the mean duration of this phase to be 265 degree-days for a set of genotypes with very variable seed weights. The difference from the value input for Afila is very small (around 1 day).

• Date of final stage in seed abortion (FSSA)

The date of FSSA depends on the total number of reproductive nodes (cf. p. 27–28) and on the rate of progression of the final stage in seed abortion along the stem, which corresponds to the fixed value of 30 degree-days between two consecutive nodes (cf. table p. 43). The number of reproductive nodes produced in non-limiting conditions is a parameter which depends on the genotype. Hence the date of FSSA can be calculated from the date of beginning of seed filling by adding the number of degree-days obtained by multiplying the number of reproductive nodes minus one by 30.

The determination of this stage is essential for the evaluation of the duration of the period "Beginning of flowering—FSSA", used for calculating the seed number (cf. p. 105). Unlike the date of beginning of seed filling, this date is extremely variable between genotypes and environmental conditions (cf. p. 27–28). The variation observed **does not** depend on the rate of progression of the stage, which is very stable for different genotypes and environmental conditions. On the other hand, it depends on the number of reproductive nodes. Consequently, it is planned to use the observed value of FSSA in the model in order to have a better simulation of the seed number.

The simulation of the number of reproductive nodes from easily-measured variables is in progress to cope with this problem, without having to provide the date of FSSA, which is not always available when the model is used. In the current version of the model, the date of FSSA is reached when the mean growth rate from the beginning of flowering begins to fall. This type of model allows the number of reproductive nodes to be predicted a posteriori from the duration of the "Beginning of seed filling—FSSA" period, divided by 30, plus one.

• Date of physiological maturity

The date of physiological maturity is very variable, depending on the environmental conditions. We can assume that it is reached when seeds cease to fill, i.e. when the maximal mean seed weight is reached (a varietal-specific parameter) or when all the remobilisable nitrogen has been remobilised from the vegetative parts (cf. p. 115, 120–121, fig. 1.44).

No effect of the environment on the maximal mean seed weight is currently simulated, in particular for situations with high temperatures or severe drought stress during seed filling. However this is at not consistent with observations made in the field.

Growth module

The growth module, based on the equation of Monteith (1977) (cf. p. 44), can be divided into three sub-modules: one for the setting up of aerial captors, one for biomass production and one for the elaboration of yield components.

• Module for setting up aerial captors

The radiation interception efficiency (RIE) depends on thermal time as described by Ney (1994). It is a logistic function that depends on cumulative degree-days counted from emergence and on the daylength at emergence (cf. p. 52–53). The parameters of the relation are independent on the crop architecture (branches, foliar type: leafless or leafed, etc.). On the other hand, this function, based solely on cumulative temperatures, is valid only for high plant densities, above 80 plants/m^2.

As this formalism is not satisfactory for autumn sowings due to a poor simulation over the winter period, another simulation has been considered. It is based on the simulation of the leaf area index from the "critical" quantity of nitrogen in the crop, i.e. the quantity of nitrogen that the crop should absorb to ensure its maximal growth (cf. p. 63–64, fig. 1.25), and to the conversion of the leaf area index into solar radiation interception efficiency.

• Module of biomass production

The radiation use efficiency (RUE) (conversion of solar radiation into biomass) varies according to the developmental stage (cf. p. 55–57); Jeuffroy and Ney, 1997; Lecoeur and Ney, 2003).

It is fixed at 2 during the vegetative period (before the beginning of flowering (BF)), then to 3,5 between the beginning of seed filling and FSSA. During the period BF-FSSA, lasting 250 degree-days, RUE varies linearly from 2 to 3,5. After FSSA, when nitrogen content in the vegetative parts is below 3 %, RUE depends on the nitrogen content of the vegetative parts calculated in the nitrogen module. The system stops (RUE = 0) when the nitrogen content of the vegetative parts reaches a limiting value corresponding to the structural nitrogen value, estimated to 1 %. In the adaptation of the model made by Vocanson (2006), the value of structural nitrogen used is 0,72 % (cf. p. 121); Lhuillier-Soundélé et al., 1999).

In the current state of knowledge, no parameter is needed to characterize the variety, either for the RUE values at each growth stage or for the structural nitrogen value. On the other hand, the effect of abiotic stresses on growth is taken into account by adding to the calculated value of RUE two functions:

— for the effect of water stress, RUE is weighted according to the variable FTSW (Field Transpirable Soil Water) (from Lecoeur and Sinclair, 1996), parameterised by Beraud (unpublished), an output variable of the water balance module described below;
— for the effect of heat stress, RUE is weighted according to an indicator RT, a function of the mean temperature (Guilioni et al., 2003);
— To take into account the effect of low temperatures on RUE, particularly during winter, the function RT proposed by Guilioni (1997) for a wider range of temperatures, is now integrated in the model.

• Module for the elaboration of yield components

This module estimates the number of seeds that are filling and calculates daily the seed demand for biomass. Hence it allows the evolution of the seed weight during seed filling and the final yield to be simulated.

The seed number is linearly related to the growth rate calculated between BF and FSSA (cf. p. 105) and fig. 2.3; Guilioni *et al.*, 2003). The conversion factor of the growth rate into seed number is 1866 (Ney and Duc, 1996).

This relationship has been validated for a very large number of environmental situations including some abiotic stresses (cf. fig. 2.3.; Guilioni *et al.*, 2003). On the other hand, for genotypes with a mean seed weight less than 200 mg, which is quite frequent in winter genotypes, the conversion factor seems to be higher, as has been observed on Frisson (Ney, personal communication).

Yield is a cumulative function of the daily demand for biomass by the seeds $(g/m^2/d)$:

$$SDB = SN \text{ filling} \times T \text{ mean} \times SGR$$

with SDB = instantaneously seed demand for biomass, SN filling = Seed number that are filling and SGR = seed growth rate.

T mean is the daily mean temperature.

In the current version of the model (Vocanson, 2006) it is assumed that between the beginning of seed filling (BSF) and FSSA, the demand for biomass by the seeds is initialized at FSSA by the calculation:

$$SDB = SN/2 \times \text{sum of degree-days from BSF to FSSA} \times SGR$$

where SN = Seed number (estimated from the growth rate between BF to FSSA). This number is divided by 2 to take into account the linear increase in the seed number from BF to FSSA and thus to not overestimate SDB at FSSA.

Seed growth rate corresponds to the individual rate of accumulation of dry matter in a seed. It is a parameter fixed at 7.10^{-4} g of dry matter per degree-day. This value has been found for cv. Solara by Lhuillier-Soundélé (1999). It can be regarded as valid for the other spring varieties with similar 1000-seed weight (250 to 300 g). It is however not suitable for small-seeded varieties (180 to 200 g 1000-seed weight). To simulate seed growth rate for a wide range of genotypes, a relationship was established between seed growth rate and seed size (Munier-Jolain, unpublished).

Seed filling stops when the maximum seed weight is reached (the seed weight, a varietal parameter, is fixed at 301 g for Solara) (cf. p. 116), i.e. when the minimum value of nitrogen content in the vegetative parts is reached.

The final yield is reached when the seed number and the single-seed weight is definitely fixed.

• Water balance module

The objective of this module is to calculate the quantity of soil water transpirable by the plant available in its rooting zone (Available Soil Water:ASW) and to calculate an indicator of drought stress (FTSW) that will affect certain simulated processes. This indicator corresponds to the ratio of the quantity of water transpirable available in the root zone (ASW) to the maximum quantity of transpirable water that the root zone can contain (TTSW).

The maximal quantity of soil water transpirable (TTSW) is defined as the product of the soil depth where roots can be observed (RSDj and the maximum quantity of water that can be extracted from the soil (RUmmax), which depends on the soil texture (Ratliff et al., 1983).

$$TTSW = RSDj - RUmmax$$

RUmmax is around 0,13 mm of water per mm of soil in almost all agricultural soils, except the most sandy and clayey. When measurements of volumetric moisture content of the soil are available, RUmmax can be calculated as the difference between the value at field capacity and the minimal value obtained after a long period without water input.

The water balance proposed works like a reservoir: the water in the different soil layers overflows into the layer below when they have reached their field capacity. This approach does not account for capillarity. In the water balance proposed in Afila, there are 2 compartments: one in the upper part that corresponds to the zone currently explored by the roots and the second one that corresponds to the lower part of the soil, that the roots may explore later, after the calculation date. When this compartment receives more water than its field capacity, drainage occurs. The input variables needed for this module are air temperature (Ta), the reference evapotranspiration (ETPref), rainfall (R) and irrigation (Water supply:WS) when necessary. The maximal soil depth explored by the roots is defined either by the physiological limit of the plant to explore the soil (Root Soil Depth) or by a physical limitation in the soil (RSDMax). The maximal rooting depth will be the lesser of the two values. The depth of the upper compartment depends only on the growth rate of the roots. The depth of the roots at emergence is fixed at 150 mm, after which it increases linearly with thermal time (RSDj = RSDj-1 + 0,7 x Tmean, see Tricot (1993)). Root growth stops when they reach the lesser value of RSD or RSDMax.

• Module of nitrogen balance

The daily nitrogen requirement of the crop is calculated from the maximum quantity of nitrogen that the plant can accumulate, as allowed by its growth in dry matter. This maximum quantity of nitrogen is estimated from the maximal curve of nitrogen content (cf. p. 65 and fig. 1.25). The daily accumulation of nitrogen in the crop is also limited by the product of mean temperature of the day and a parameter: SNAmax: Maximal rate of nitrogen accumulation (kg/ha/°C.d) by the crop, estimated at 0,77 kg.ha^{-1}.°C.d^{-1} by Maltas (2002).

As it has been shown before (cf. p. 65, 74–75), peas accumulate nitrogen by symbiotic fixation and by absorption. The quantity of nitrogen fixed daily depends on the nitrogen availability in the ploughed area: Nmin (d) estimated by the module of soil nitrogen mineralization (cf. § below) (Voisin, 2002):

— before beginning of flowering: %N fixed = –2.96 × Nmin(d) + 142.77
— after beginning of flowering: %N fixed = –3.27 × Nmin(d) + 110.19

It depends also on the development stage of the crop (Voisin, 2002).

When the sum of degree.days from emergence is less than 235°C.d, there is no fixation. Before seeds start to fill, the quantity of fixed nitrogen is calculated as:

Nfixmax(d) = (–0.0000357 × stl(d) + (0.0119 – (–0.0000357 × (sdd (emergence) – df + stdf – drgph))) × DM(d)

Before the final stage in seed abortion (FSSA) is reached, the estimate can be made with: Nfixmax(d) = 0.0119 x DM(d)

After this stage, it is assumed that fixation stops.

The rate of nitrogen fixation is the lower result from the two calculations given above. We would like to add a limitation to fixation due to excess water, as may happen with autumn sowings, or with water stress from the beginning of flowering, causing an irreversible cessation of fixation. However, as the model stands, the quantity of nitrogen accumulated by a crop allows its state of nitrogen nutrition to be defined, in the absence of water stress, by the Nitrogen Nutrition Index (NNI) (cf. p. 157–158). This index is then used to simulate the effect of a nitrogen stress on the daily dry matter production (cf. p. 159).

This module also relies on a calculation of the quantity of mineral nitrogen available in the soil every day. This is obtained from a module of soil nitrogen mineralization in the model AZODYN (Jeuffroy and Recous, 1999). It simulates the daily quantity of mineral nitrogen available in the rooted soil layer depending on the physico-chemical characteristics in the soil, the previous crop, the nitrogen residues at sowing, the mean temperature and the possible leaching of nitrates towards the unrooted layer. Finally, the quantity of nitrogen leached during the whole growing period is estimated.

• Module for partitioning nitrogen within the plant

This module drives crop behaviour after the beginning of seed filling. It is based on the comparison of the nitrogen supply in the plant with the seed demand for nitrogen.

The cumulative demand for nitrogen by the seeds is estimated from the cumulative Demand of Seeds for Biomass (DSB) and their nitrogen content, which is assumed constant during seed filling. This has been arbitrarily fixed at 4 %. However the results of Lhuillier-Soundélé (1999) on many cultivars subjected to different nitrogen regimes indicate that environmental variability exists for this parameter (1.82 to 4.53 %). Likewise Burstin *et al.* (2003) have demonstrated that considerable genetic variability exists for this parameter, currently not taken into account because the genetic diversity in cultivated varieties is very small.

In the current version of the model, as for the cumulative seed demand for biomass, the seed demand for nitrogen is estimated from the FSSA. The daily demand of seeds for nitrogen depends on the daily mean temperature, on the rate of nitrogen accumulation in seeds (a genetic parameter depending on the maximum seed weight) and on the number of filling seeds (all after FSSA). The initialization at FSSA of the cumulative demand of seeds for nitrogen between BSF to FSSA is estimated as the product of mean nitrogen content of seeds at FSSA (estimated from results of Lhuillier-Soundélé, 1999), the number of seeds at FSSA and the mean seed weight at this stage.

The supply of nitrogen within the plant is made up of the uptake of exogenous nitrogen (external supply) and transfer from the remobilisable pool in the vegetative parts (internal supply), (Larmure and Munier-Jolain, 2004):

— The internal supply is fixed at the beginning of seed filling (BSF). The nitrogen content at BSF (Ntot) is calculated using the critical dilution curve (cf. p. 62) from the dry weight simulated at BSF (DMBSF). The quantity of nitrogen accumulated until BSF and available for internal transfer is: Ntot × 0.01 × DMBSF;

— The quantity of exogenous nitrogen (= quantity of nitrogen accumulated after BSF) on day d is estimated as the product of Ntot and the dry weight simulated on day d, as it has been shown that crop nitrogen content is relatively constant from BSF (cf. p. 83, Ney *et al.*, 1997). No immobilization of exogenous nitrogen in the immobile (structural) nitrogen pool is currently taken into account. This could lead to an overestimation of the quantity of nitrogen available for seeds if vegetative growth is maintained during seed filling.

The comparison of the supply and demand for nitrogen determines the quantity of nitrogen remaining in the vegetative parts. This output variable allows RUE to be predicted at the end of growth (as described before).

From FSSA, the supply of nitrogen is estimated every day as the total quantity of nitrogen accumulated in the crop (nitrogen accumulation module) minus the quantity of nitrogen already accumulated in the seeds and the nitrogen unavailable to the seeds, estimated at 0.72 x DMBSF (immobile structural nitrogen).

Prospects

The model in Afila which includes water and temperature stress has been validated on a large database. The result is that simulation of growth and yield is good when severe water and high temperatures stresses occur. For less severe stresses, simulation is less good. In very favourable conditions, the model is unsatisfactory and it is not possible to simulate very high levels of biomass and yield. This is partly because nitrogen in the vegetative parts is used too quickly in the model and the simulated lifespan is too short. Two parameters influence the nitrogen content of the vegetative parts and hence the lifespan: the nitrogen content of the seeds and the maximum value of radiation use efficiency (RUE). Sensitivity tests have been carried out using many values for these 2 parameters. The result is that the combination of values of parameters which minimises the differences between simulated and measured biomass is Ngr = 3 % and RUEmax = 4. These values also agree with measured values.

The modifications made to the initial model to enlarge its scope to include autumn sowings and photoperiod-sensitive genotypes and to take into account limiting or luxury nitrogen nutrition may be evaluated on many data sets that we are acquiring. These include a module to simulate the effect of the soil structure on mineralization, growth and finally on yield and quality elaboration processes, which is at present being refined.

This module seems necessary to account for differences in behaviour between spring and autumn sowings, for which moisture conditions at sowing are very different and can have major consequences for the subsequent growth of this crop, which is particularly sensitive to this limiting factor (Cf. p. 162–163).

Finally, it is planned to introduce other stress modules and, in particular, biotic stresses, which also need to be considered to simulate the different behaviour of spring and autumn sowings. For example, the work on Aschochyta (Cf. p. 182 to 193) and, in particular, the effect of the disease on the plant growth could be integrated further into the global model.

9

Proposal for a diagnostic approach to analyse yield variations in peas

Isabelle Chaillet, Véronique Biarnès

Objective and general presentation

For the pea crop there is considerable variability in yield between years and regions, and within a small region for a given year (Doré, 1992). To understand this variation and to identify its causes, the crop diagnosis method has been proposed and implemented for various species (Doré *et al.*, 1998). Crop diagnosis enables one to identify and rank the limiting factors which reduce yield by distinguishing the effects of soil and weather from those of cultural practices, and to identify solutions to put into practice to limit the occurrence of these factors.

Description of the diagnostic approach

The diagnostic approach is made up of three steps :

- The environment is characterised using appropriate variables, such as the physico-chemical characteristics of the soil, weather variables, environmental conditions created by cultural practices in interaction with crop behaviour (for example by disease assessment);
- Next, the affected yield components have to be identified. For this, the level of each component is estimated by comparing it with previously established reference curves which characterise the physiology of a given variety in the absence of any limiting factors. All the knowledge presented in parts I and II consists of elements which can suitably describe the key stages of crop growth;

- Lastly, on the basis of knowledge acquired and the available modelling tools, (cf. part III.1.), the periods within the growth cycle during which these components are formed are precisely determined. If a component has a low value compared with its reference value, one has to look for the factors which may have limited its elaboration during its period of formation. Table 3.1 summarises the main limiting factors known for all of the components.
- To identify the affected components, one begins by drawing graphs representing yield as a function of grains/m^2, the yield as a function of single grain weight, the number of grains/m^2 as a function of the number of plants/m^2 etc. Then, to find the factors which may have reduced a component, one can draw graphs relating certain components to the weather variables which seem to be the most appropriate.

The use of a crop model can serve as a reference for diagnosis. On the one hand, although it does not include the limiting factors responsible for the low yields observed, it serves as a reference against which the observed yields can be compared. Thus, a yield lower than that predicted by AFILA (cf. part III.1.) may be due, for example, to nitrogen deficiency or health problems, such as pests and diseases, as these factors are not covered by the model. In the case of root diseases (*Aphanomyces*), the model enables the difference in yield between the potential allowed by the climate and that obtained in the presence of the disease to be measured (Alamie, 2002). On the other hand, if it already includes certain limiting factors (as in the case of the model AFILA and weather factors), the comparison with simulations carried out using other weather scenarios may lead to a resolution of the nature of the limiting factors responsible.

Examples of the use of the diagnostic approach

The diagnostic approach can be used :

- *to carry out a seasonal appraisal each year*: in the different French production zones, the factors limiting the yields obtained are identified. For those involved in the protein industry (producers, breeders, technical advisors, research organisations etc.) these retrospective diagnostic elements can be used to re-evaluate the appropriateness of the practices and technical advice currently in use, to provide useful information for improving varietal types (the ideotype concept) and finally, to validate the knowledge acquired about the physiology of the crop, and suggest new hypotheses.
- *to rank the factors limiting pea yield in a given region*, with a view (i) to direct studies which could be carried out later in cases where

Table 3.1. Factors limiting yield components.

Yield component	Abioitic stress	Biotic stress	Integrative variable
Seeds sown per m²	-Precision of the sowing machine -Adjustment of the sowing machine		
Low plants per m² at emergence	-Frost during imbibition -Dry soil in the first 5 cm -Excessive water in the soil : rot seed -Crust at the surface of the soil	-Sowing smelt -Toxicity of treatment against weeds -pigeons, rabbits,....	
Plant losses during the winter	-Frost damages -Excessive water		
Low number of branches per stem	-Too deep sowing (prevent the emission of branches)	-Toxicity of treatment against weeds	Low crop growth rate (See table 3.2)
Low number of fertile stems per m²	-Frost after floral intiation (frosten apices)	-Thrips	Low branches per stem (See above)
Low number of flowering nodes	-Water stress during flowering -High temperatures before flowering	-Aphanomyces -Thrips at the beginning of the cycle -Strong attacks of aphids	-Nitrogen deficiency before beginning of flowering and which carry on (See table 3.2)
Very high number of flowering nodes	For indeterminated cultivars, in case of of very high water supplies during flowering		
Low seed number per stem		-Aphids -Midges -Ascochyta blight -Weeds	-Low number of flowering nodes (See above) -Low crop growth rate (See table 3.2)
Low seed number per m²			-Low crop growth rate from BF to FFSA (See table 3.2) -penality for one or moe yield component which participates to its formation
Low thousand seed weight	-Very high seed number per m² -Water stress -High temperatures -Compacted soil structure	-Aphids -Bruchids -Pea moth larvae -Ascochyta blight	
Low protein concentration	-Compacted soil structure		-Low availability of nitrogen assimilate (nitrogen deficiency and/or very high seed number)

these factors and their effects on the physiology of the crop are not well understood and/or the technical solutions to remedy them do not exist or are not yet perfected, or (ii) to test the value of the crop in regions where it is currently rarely grown in order to identify the technical problems preventing its introduction. Ranking the limiting factors by region is an important element in the concept of pea ideotypes and when taking account of « genotype x environment » interactions in the choice of variety in a given environment well characterised by a field of constraints. This last point is also applicable to test the value of other legumes in French or European regions where they are not yet grown. Thus, this approach was used from 2001 to 2004 to test the suitability of various autumn- and spring-sown protein crops to the Pays de la Loire region.

- *to quantify the loss of yield associated with limiting factors for which it is not possible to make experimental comparisons* since there is no means of controlling these factors. For example, during work on late season root diseases carried out in Normandy in 2004, the model made it possible to determine the difference in yield between the potential allowed by the climate and the yield obtained in the presence of the soil fungi responsible for the root diseases observed at crop maturity.

Characterisation of the environment

The variables which characterise the environment can be of various kinds :

- Climatic (radiation, temperature, water availability etc.);
- Biological (presence of pests, diseases, weeds etc.);
- Physical (soil structure);
- Chemical (availability of mineral elements).

Climatic factors

Different climatic variables, calculated from meteorological data obtained from weather stations close to the studied fields can be used to establish relevant indicators of the weather during key periods in the elaboration of yield components. Pea growth is usually sub-divided into three phases:

- a phase of canopy growth, from emergence until the beginning of flowering, corresponding to the installation of the structures for capturing radiation and mineral resources. In fact, at the beginning of flowering, radiation interception should be about 1 if optimum growth is to be achieved in the following phase (cf. part I.2.);

- a phase of grain formation, from the beginning of flowering to the final stage in seed abortion (« BF-FSSA »). The growth rate of the stand during this period determines the number of grains formed. (cf. part on the seed number);
- a grain-filling phase, from FSSA until physiological maturity, partly determined by the component «mean single grain weight » (cf. part on individual seed weight).

It is important to understand the timing of developmental stages to calculate the weather variables. The beginning of the flowering stage is usually observed, but it can be simulated (cf. part on the floral initiation and the beginning of flowering). The FSSA stage is generally simulated (cf. part on reproductive development). Physiological maturity is observed or simulated (cf. previous part on a model which integrates knowledge on pea crop physiology and agronomic diagnosis). Hence for the determination of the development stages punctuating the key stages of yield elaboration, modelling tools are crucial as they make it possible to avoid a large number of measurements of development which are essential for the diagnosis, but often too troublesome to implement routinely in the field.

• Photothermic quotient

The yield is determined mainly by the number of grains formed, which is highly correlated with the growth rate during the « BF-FSSA » phase. During this period, the growth rate is influenced by the radiation (the driving force for growth) and temperature, which has a big influence on the length in days of this period. The photothermic quotient can therefore be useful for describing the weather conditions during seed formation. It is the ratio of the daily radiation sum during the BF-FSSA period to the mean daily temperature sum during this same period.

• Study of cold temperatures

To assess frost damage to winter peas, percentage losses of plants can be calculated provided that the plants were counted at emergence. Apart from the loss of plants, an estimate of the frost damage to the remaining plants can also be made using the frost damage scores shown in part II p. 180.

For apical frost, which destroys the cauline meristem of the stem in autumn sowings but also in spring sowings due to late spring frosts, the temperature threshold has not been fixed with certainty. It seems to be in the order of $-2°C$ to $-4°C$ (Mangin, 1998). It is possible to calculate a cumulative minimal temperature below $-4°C$ (between sowing and the start of flowering) for comparing sites. This variable, used by Courty (1999) in a network of spring pea varieties planted in several European countries, revealed one site where late frosts were particularly frequent.

- ## Study of high temperatures

For high temperatures, a cumulative maximum temperature above 25°C has proved useful to detect periods of high temperature during the three key periods of growth previously described. Thus, whenever they occur during growth, temperatures above 25°C reduce the efficiency of conversion of radiation into biomass (cf. part on effects of high temperature on a pea crop). During the BF-FSSA period, high temperatures reduce the number of seeds formed when this accumulation is more than 10°C (Jeuffroy, 1991). For the other phases, no threshold value has been determined. Before the beginning of flowering, high temperatures lead to a reduction in leaf area and in the number of reproductive nodes formed during flowering. In view of the diversity of climate in the French pea-growing regions, the frequency and timing of heat stress events vary between regions.

- ## Water balance

A first method consists of representing changes in the amount of soil moisture reserve as a function of calendar dates. Then one has to locate the moment at which water stress occurred, i.e. to determine the dates when the level of the reserve falls below 2/3 of the available water capacity. At this moment, the readily available water is exhausted; the plant has used all the available water and begins to suffer seriously from lack of water (water stress; cf. part on influence of water deficit on pea canopy functioning).

A second, more synthetic method is based on the calculation of the ETR(Real Evapotranspiration)/ETM (Maximal Evapotranspiration) ratio for each phase of pea growth. Below a value of 0.7 for this ratio, the plants begin to suffer from lack of water, and the yield begins to be affected.

A third way to characterise the moisture state of the soil is to use the variable FTSW; it is in fact possible to predict the reaction of the plant according to the value taken by this variable (cf. part on influence of water deficit on pea canopy functioning). Situations can be classified in relation to one another using the values of FTSW.

Biological factors (presence of pests, diseases etc.)

Only observations of plots can establish whether parasites are present in the crop. For example records of leaf damage by weevils can be useful for deciding the level of attack of the crop by this pest and the consequences for the nitrogen nutrition of the peas (cf. part on sitona p. 193). A record of the presence of anthracnose (cf. part on impact of ascochyta blight on spring pea functioning yield p. 187) can also enable the effects of the disease on plant growth to be measured.

Physical factors (soil structure)

An observation of the cultivated soil profile (Manichon and Gautronneau, 1987) is necessary to judge the structural state of the soil. The variable « percentage of delta clods » can be used to judge this state and appears to be a good indicator of the reduction of the biomass produced by the crop(cf. part on effects of compacted soil structure).

Chemical factors

An analysis of the mineral element content of the soil can reveal possible deficiencies.

Study of yield elaboration

Different variables can characterise the yield elaboration of the pea crop. Some allow a diagnosis to be made during growth, such as measurements of nitrogen content and biomass at the beginning of flowering. Others, such as a study of yield components, can provide a retrospective appraisal.

Assessment at the beginning of flowering

To carry out an assessment at the beginning of flowering is an important element in pea crop diagnosis. In fact, from the beginning of flowering, peas should have maximal growth until FSSA in order to form the largest possible number of seeds (cf. part on the seed number). For this therefore it is necessary that by the beginning of flowering, the crop should have reached a maximal level of radiation interception (approaching 1) if it is hoped to obtain a high yield (cf. part on carbon acquisition at the crop level in pea).

Measurements of biomass and nitrogen content at the beginning of flowering can indicate the crop potential at this stage. If the level of biomass and/or nitrogen in the plants are below the potential established from reference curves, it means that a problem has occurred between sowing and the beginning of flowering and the yield potential may already have been compromised.

For the biomass, it is generally considered that a pea field should have reached a threshold of 280–300 g DM/m² by the beginning of flowering. Causes for low biomass at this stage are given in table 3.2.

The nitrogen nutrition index (NNI) (ratio of the measured nitrogen content to the critical nitrogen content) is a good indicator of the level of nitrogen nutrition of a pea crop (cf. part on nitrogen deficiency, p. 157–158):

Table 3.2. Limiting factors affecting the functionning of the plant.

Component	Aboitic stresses	Biotic stresses	Integative variable
Low branching	-Compacted soil structure -Late sowings		
Nitrogen deficiency	-Water Stress -Excessive water -P and K deficiency -Compacted soil structure -Low availability of nitrogen in the soil (low mineralisation, drought,...)	-Strong attacks of pea weevil (destruction of nodules) -Aphanomyces -Weeds	-Low nodule biomass
Low biomass at BF	-Water stress before BF -Excessive water (asphyxied plants) -Compacted soil structure -Late sowings (1)	-Aphanomyces -Competition with weeds -Ascochyta blight	-Nitrogen deficiency before BF
Low crop growth rate from BF to FFSA	-Low radiation -Low ratio of adiation and temprature (for example in case of heat time), the growth is insufficient to produce lots of seeds -Water stress -High temperatures	-Aphanomyces -Competition with weeds -Ascochyta blight	-Nitrogen deficiency (See above)
Végétation excessive	For indeterminated cultivars, in case of very high water supplies during all the cycle : production of vegetative parts to the detriment of seeds		

BF : Beginning of flowering
FFSA : Final stage in seed abortion
(1) : not enough days from sowing to beginning of flowering, then low sum of radiation between this two stages
Warning : This table is not complete. The objectif is to try to list the main factors affecting the functionning of the plant to help the reader to make a diagnostic in pea fields. There is no hierarchy in the presentation of the different factors.

- If it is 1 or more (i.e. the observed points are situated on or above the dilution curve) it means that nitrogen nutrition has not limited growth;
- If it is below 1 it means that there is nitrogen deficiency (cf. part on nitrogen deficiency). The yield potential at the beginning of flowering is therefore reduced (Ney et al., 1997).
- Nitrogen nutrition has important repercussions on the plant structure. Thus, in numerous situations where there is nitrogen deficiency before the start of flowering, the plants have reduced growth and a colour ranging from pale green to yellow, a sign of a deficiency of the photosynthetic apparatus : they have therefore a stunted and feebly branched structure, as the number of branches is greatly influenced by the growth rate of the plant (cf. part on branch emergence). These plants later develop a small number of reproductive nodes and a small number of seeds (Sagan et al., 1993), so that the yield suffers. Conversely, in numerous situations the NNI observed at the beginning of flowering is in the order of 1.1 to 1.4; the peas have therefore accumulated more nitrogen than is strictly necessary to assure maximum growth. These situations, which are extremely « prolific » and favourable at the beginning of growth (well-watered), lead to highly developed and very branched plants.

Study of yield components

Yield can be analysed as the product of the number of seeds per m^2 and the weight per grain. In general, the component « number of seeds per m^2 » is the one which explains most of the yield variation (Doré et al., 1998). The number of grains per m^2 can itself be expressed as the product of two components—the number of fertile stems per m^2 (measured) and the number of grains per stem (obtained by calculation). Note that there is a certain capacity for compensation between these two components : when the number of fertile stems is very high, the number of seeds per stem is low. Conversely, when the number of fertile stems per m^2 is low, for example following poor emergence, the plants compensate at least partially by producing a large number of seeds per stem. The component "number of fertile stems per m^2" is the product of the number of plants per m^2 emerged (or at the end of winter for winter peas) and the number of fertile stems per plant.

The mean yield components (number of seeds per m^2, 1000 seed weight, number of fertile stems etc.) depend on the variety. In a diagnostic approach, it is preferable to avoid the variety effect and work with the same variety in all the fields, using a control variety or standard genotype (Brancourt-Hulmel *et al.*, 1999). Also, it is preferable to choose a known variety, for which data exist as regards the yield components to reach.

• Number of seeds/m²

As there is a linear relationship between the number of seeds per m² and the growth rate from the beginning of flowering until the FSSA, all the factors affecting growth rate during this period have repercussions on the number of seeds per m² established. These are summarised in table 3.2.

• Seed profile and simplified profile

The grain profile (distribution of seeds over the different reproductive nodes) is not strictly a yield component, but it constitutes a useful indicator of the way in which the number of seeds per m² is achieved (cf. part on the seed number). The « grain profile » graph represents the number of seeds per node for each fruit-bearing node number (cf. methodology file).

According to Jeuffroy (1991), one finds essentially two kinds of profile, illustrated in figure 3.1 : « low and pot-bellied » and « high, with chest thrust out » profiles. There is a direct relationship between the seed profile and the vertical structure (or simplified profile) which has the advantage of being much more quickly measured. The measurement of the vertical structure gives the number of the first fruit-bearing node and the number of the last fruit-bearing node (cf. methodology file). If the number of the first fruit-

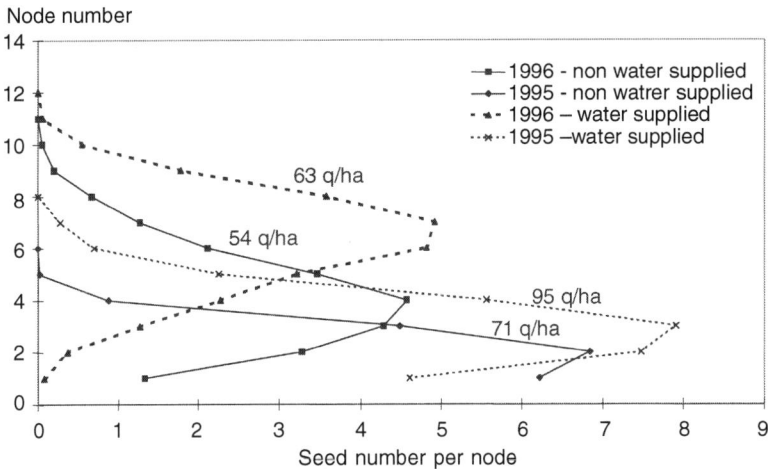

Figure 3.1. Example of seed profiles. Situations in 1995 had a « low and pod-bellied » profile, situations in 1996 had a « high, with chest thrust out » profile. Cv. Baccara, Le Magneraud (17) in 1995 and 1996. (From Gaillard and Bouthier, Perspectives Agricoles n°220, January 1997) .

bearing stage is close to 1 and that of the last is between 3 and 4, the profile is short, corresponding to a « low, pot-bellied » seed profile. If the number of the first fruit-bearing node is close to 2 and that of the last fruit-bearing node is more than 5–6, the profile is high, corresponding to a « high, chest thrust out » seed profile.

In the example shown in figure 3.1, the yields obtained with only 4–5 fruit-bearing nodes are higher than those obtained with 8–9 such nodes. In fact, for cultivated grain legumes varieties, a very large number of stages is often associated with excessive vegetative growth at the expense of seed production, and hence a severe fall in yield and a large mass of vegetation. In the nineties, to reduce the risk of having excessive vegetation at the expense of seed (a low harvest index), plant breeders aimed to produce semi-determinate varieties, a characteristic closely associated with an increase in the 1000-grain weight. This progress in selection was possible partly because of the improved understanding of pea physiology.

If the number of the last fruit-bearing node is high (greater than 7 for spring peas), it means that the crop has been well supplied with water throughout flowering. Conversely, a small number of fruit-bearing nodes (3–4 in spring peas) could have several causes (tables 3.1 and 3.2) for example :

- Moisture stress during flowering, with a ETR/ETM threshold below 0.7 or when the readily available soil water is exhausted. This is most often the case;
- High temperatures before the beginning of flowering. This phenomenon is rather rare;
- Nitrogen deficiency. This can occur in the event of drying out of the seedlings at the start of flowering, soil compaction or weevil attack etc.

In a cropping situation, the analysis of the seed profile in comparison with available reference data or with a simulated profile (cf. part on the seed number, p. 108–109 and fig. 1.39) can reveal problems which occurred at a given moment during seed formation such as a gall midge attack, brief high temperatures, a short frost etc.

- ## 1000 seed weight

A low 1000 seed weight may have a number of causes (tables 3.1 and 3.2).

- A very large number of seeds per m^2 can result in poor seed filling if growing conditions do not allow enough available assimilates during seed filling;
- Moisture shortage towards the end of growth reduces the 1000 seed weight if the ETR/ETM ratio during the filling stage is below a value between 0.3 and 0.5. For example, an analysis of limiting factors in the central region (Arvalis—Plant Institute, 1996) shows that one

can classify fields into two groups according to the level of the ETR/ETM of FSSA at maturity : below about 0.5, the 1000 seed weight is below 300g and averages 280g. On the other hand above 0.5 it is above 300g, averaging 330g. This trend was also observed in the network studied in 1995 in the central region;

- High temperatures during filling seem to reduce the seed weight (cf. part on the individual seed weight). Thus, in the irrigated Arvalis Plant Institute trials in the Rhone Alps, the 1000 seed weight tended to be lower in years when high temperatures were observed after FSSA.

However we should remember that the 1000 seed weight component does not depend simply on the environmental conditions during seed filling. In fact the growth potential of the seed during this period depends on the environmental conditions during the « BF-FSSA » phase (cf. part on the individual seed weight).

• Protein concentration

Environmental conditions usually explain most of the variation in pea grain protein concentration (cf. part on the seed protein concentration and part on model). There is much variability between sites in a given year, but no particular type of land can be said to exhibit stability in this respect, and the ranking of soil/climatic zones is not the same from year to year (Carrouée and Duchêne, 1993). For a given variety, the degree of variation can be very high—more than 10 points (Carrouée and Duchêne, 1993). The effect of environmental conditions on the protein concentration of pea seeds has been studied by various authors. The results are sometimes contradictory (Biarnès, 2000).

The protein concentration at harvest may be related to the nitrogen nutrition index (NNI) at the beginning of seed filling :

- A low NNI at the beginning of flowering or of seed filling (BSF) leads very often to a low seed protein concentration at harvest. Thus, in comparisons between compacted and uncompacted soils there was a positive relation over all the study years between the NNI at DRG and the protein content at harvest (cf. part on effects of compacted soil structure);
- A NNI at the beginning of flowering above 1 often leads to a high protein content, but not invariably. It depends on the number of seeds formed and the weather conditions after the beginning of flowering. Favourable conditions for yield are also often beneficial for protein concentration. It is in fact possible to accumulate a high yield with a protein concentration which is also high. This was observed in the pea production areas of northern France in 1999. Fields of peas giving

yields of about 7500–8000 kg/ha resulting from a high seed number (>2600) and a high 1000 seed weight (285g) also had high protein concentrations (25 % of dry matter). In that year the weather conditions were very favourable throughout the growing period (high radiation and mild temperatures during seed formation, and rain and moderate temperatures during seed filling).

- Harvest index

The harvest index is not itself a component but a variable which can characterise the behaviour of the crop retrospectively (cf. part on biomass and N partitioning during crop growth).

The harvest index (dry weight of seeds/total dry weight of seeds + vegetative parts) can illustrate the problem of transfer of assimilates from the vegetative parts to the seed, as was shown earlier for a large number of flower-bearing nodes. Apart from an excess of vegetation (for example where irrigation has been excessive and too early on an indeterminate variety), problems of transfer of assimilates can be associated with biotic stresses interfering with the transfer of assimilates within the plant (e.g. anthracnose, cf. part on ascochyte blight).

- Similar yields with very different plant architecture

It is possible to obtain the same yield with very different architecture. For example a yield of 5000 kg/ha can be obtained in quite different situations :

- with severe water stress from the beginning to mid-flowering : there can be only 3 floral nodesformed, giving 3 fruit-bearing nodes correctly filled with seeds;
- in a situation with excessive water during flowering, with early heavy rain impairing the plant nutrition, there may be 8–9 floral nodes formed, with many aborted seeds, leading to a small number of grains per stem.

In both situations 120 fertile stems per m², a 1000 seed weight of 250 g, and 2000 seeds per m² are produced. In the first case, each nodebears on average 5.5 seeds (low pot-bellied profile) with very efficient use of the dry matter produced (i.e. a high harvest index). In the second case, there are only 2.5 seeds per node on average (high profile) : the harvest index is therefore low, as an excess of vegetative dry matter has been produced at the expense of seed.

- Use of modelling to improve cultural practices

During a study made in the Sarthe in 1995 and 1996 on a total of 23 spring-sown pea fields over the two years, it was possible to relate the observed number of grains per m² to the ETR/ETM ratio during the period

« BF-FSSA » ($r^2 = 0.82$) (Biarnès-Dumoulin *et al.*, 1998). To improve the water balance of the crop, two aspects were analysed by simulation over 22 years using a model which takes account of radiation and mean temperature only—the effect of early sowing and of irrigation practice.

- Early sowing (mid-February instead of 5 March) brings forward the date of the beginning of flowering by about one week and reduces the exposure to high temperatures during the « BF-FSSA » phase. Radiation interception is greater and the number of grains per m^2 increases by 10 %. This result justifies sowing earlier every year;
- For irrigation, for 15 of the 22 years studied, an average application of 50 mm of water to achieve a ETR/ETM ratio of 80 % during the « BF-FSSA » phase increases the yield by 1200–1300 kg/ha.

This example illustrates the value of using a model to improve cultural practices (determination of the optimal sowing date and a suggestion for managing irrigation).

Example of the use of tools to characterise the environment and the crop: classification of environments and seasons in France and in Europe

The different variables which can characterise environments (photothermic quotient, high temperatures, ETR/ETM) or the crop (NNI at the beginning of flowering) can be used to rank situations against one another. The following paragraphs indicate the range of variation obtained for these different variables in the French seasonal assessments made on spring-sown peas in recent years. An analysis on the European scale is also presented.

Towards a French regional classification

A study of the photothermic quotient and high temperatures over the last five years allows us to suggest a hypothesis about the adaptation of spring peas in the different French regions and the yield potential allowed by the climatic conditions.

- Photothermic quotient

For France, the values generally obtained in the seasonal assessments made between 1999 and 2003 for the photothermic quotient fall between 0.92 and 1.56:

- Values between 1.40 and 1.56 are obtained when the radiation sums are high and the temperature sums are average, corresponding to high radiation and moderate temperatures over a period which can be quite long. This is what is often seen in Picardy.
- The lowest values (0.9–1) are observed when the radiation is high and so are the temperatures. Usually this corresponds to a BF-FSSA period which is short because of the high temperatures, as is often found in the south or centre of France.
- The commonest mean values fall between 1.1 and 1.3.

- ## High temperatures

Table 3.3, which shows the ranges of variation encountered in the seasonal assessments made in France between 1999 and 2003 for cumulative maximal temperatures above 25°C between BF and FSSA , shows that:

- In the north of France, in Picardy, Normandy and Brittany, we observed hardly any limiting high temperatures (above 25°C) between BF and FSSA. Only the year 2000 had a cumulative temperature (above the 25°C base) of 10°C. Thus these regions only rarely (1 year in the 5 studied) experience conditions of limiting heat stress.
- In Champagne however, the accumulations observed were very often high (more than 20°C for 4 of the 5 studied years). The temperature therefore is often detrimental during grain formation in this region.
- In the Rhone Alps and in the south-west, the maximum temperatures are also often high during the BF-FSSA period and the cumulative temperatures observed exceeded the threshold of 10°C for at least 4 of the 5 years studied. Note that the highest yield at Etoile under irrigated conditions was obtained in the absence of high temperatures in 2002.

Table 3.3. Sum of maximal temperatures above 25°C from beginning of flowering (BF) to final stage in seed abortion (FFSA) for spring pea in experimental trials UNIP-ARVALIS-Institut du Végétal-FNAMS. Calculations have been based on the most used varieties. (Stage BF has been observed in fields, stage FFSA has been calculated taking into account the real number of reproductive nodes).

	1999	2000	2001	2002	2003
Vraignes (80)	3	18	3	1	14
Rots (14)	5	19	2	4	7
Bignan (56)	4	16	5	5	9
Vraux (51)	21	29	25	4	26
Le Magneraud (17)	9	10	13	10	4
Ouzouer (41)	14	24	36	12	14
Bourges (18)	20	42			
Lyon St Exupéry (69)	33	16	24	12	11
Etoile (26)	37	21	22	5	12
Bazière (31)	21	12	27	4	12

- In the west, the maximum cumulative temperatures often approach the threshold of 10°C and only exceeded it a little in a single year (2001). This perhaps explains why the yields obtained at this site with irrigation were often higher than those observed in the Rhone Alps, also with irrigation.
- Finally, in the centre, high temperatures often appear to be detrimental during grain formation. The accumulations are clearly above 10°C in at least two of the 5 studied years for the two sites considered (Ouzouer, 41 and Bourges, 18). This factor therefore is often limiting for grain number.

• An analysis of yield potential

Hence the north of France (Picardy, Normandy and Brittany) has the most favourable climatic conditions as regards the ratio of radiation to temperature. Excessive temperatures are unusual there. This undoubtedly explains why in these regions one observes the highest yields almost every year (7000–8000 kg/ha for the best fields). Furthermore, the soils in these regions are mostly deep, with a large water holding capacity, and the rainfall is also abundant, limiting the risks of water stress. Central France, on the other hand, is a region where high temperatures are frequently harmful during grain formation. The best yields with irrigation are usually between 6000 and 7000 kg/ha.

In the west, the growing season, often ahead of other regions to the east and north due to early sowing, manages to escape the high temperatures. Irrigation, when it is possible, makes up for the shortage of rain often found in this region where the soils generally have a low water holding capacity. With irrigation one can reach 7000 kg/ha.

Finally in the Rhone Alps, in spite of irrigation, which is essential, high temperatures act at two levels—on the one hand they shorten the life cycle and lower the radiation/temperature ratio; on the other, they result in a fall in the number of seeds/m² produced. This undoubtedly explains why, even with irrigation, yields in these regions are limited to about 5000 kg/ha in very hot years.

Diagnosis therefore enables us to rank the limiting factors in the different pea production zones and to better understand the different production levels obtained in each of them. It is however necessary to confirm these results over many years to be able to generalise about the trends observed. For the regions likely to suffer high temperatures early in the pea growth cycle (Champagne, centre, south-east and south-west), growing winter peas, which develop earlier than those sown in spring, could be a way of avoiding these climatic stresses towards the end of growth.

An example of comparison between years: assessment at the beginning of flowering

The analysis of values of biomass and nitrogen concentrations observed at BF between 1995 and 2003 makes it possible to compare year with year. The favourable years tend to have points situated above the dilution curve. The difficult years have points below the curve. For simplicity we only show 3 years (fig. 3.2).

- In 1999, a favourable year in terms of weather (high radiation, moderate temperatures, regular rain) throughout the growth of the peas, with no serious pests or diseases, we observe high nitrogen levels at all the sites studied. The yields were in general very good.
- In 2001, there was a series of sowings : early sowing at the end of February and late sowings in April. We observe two series of points corresponding to these two periods of sowing. The points above the dilution curve correspond to early sowings and the points under the curve to late sowings. The early sowings had yields distinctly higher than late sowings. The latter suffered from unfavourable growing conditions (serious drought accompanied by high temperatures) between emergence and BF, while the early sowings enjoyed more favourable conditions during this period.

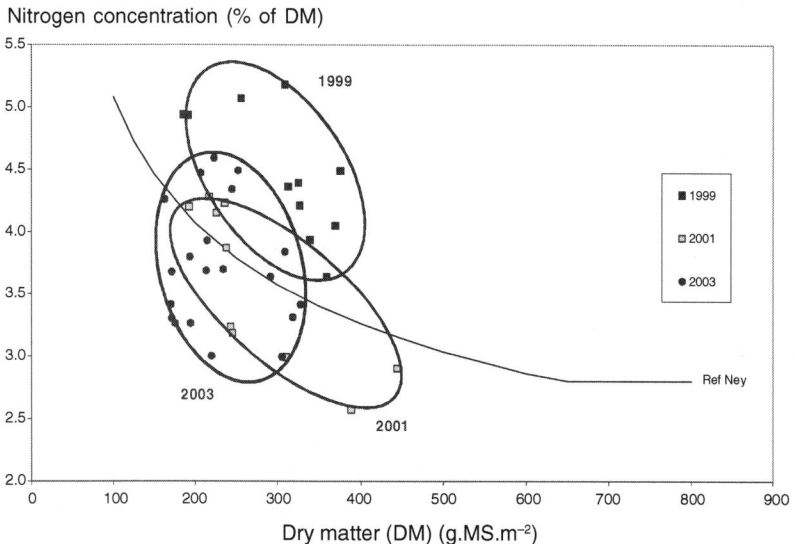

Figure 3.2. Nitrogen Nutrition Index at beginning of flowering in 1999, 2001 and 2003— All regions of France (data Arvalis—Institut du Végétal, FNAMS).

- In 2003, a year with an early drought from emergence to BF followed by rain, most of the points lie below the theoretical curve. Just a few points are slightly above the curve. They correspond to fields which were irrigated early on, during emergence or just before BF, or to some rare situations on deep soils which received rain during this period. The disappointing yields in 2003, despite the good appearance of the crop common at flowering due to the return of favourable conditions (rain) after BF, is largely explained by nitrogen deficiency before the start of flowering.

Take the particular case of chalk soils. In every year except 1999, on such land the NNI was always below 1. On these soils, plant growth is worse than on silty soils, at least until BF, and the plants are usually poorly branched.

• Classification of production zones atthe European scale

At the European level, a study on many years was made on spring-sown peas using the following weather variables :

- minimum temperature sums below –4°C between sowing and BF
- maximum temperature sums above 25°C from 9 leaves to BF, from BF to FSSA and from FSSA to PM;
- sums of ETR/ETM ratios below 0.7 from 9 leaves to BF, from BF to FSSA and from FSSA to PM;
- rainfall totals between Physiological Maturity (PM) and PM + 10 days;
- biomass predicted by a growth model taking account only of radiation and mean temperature.

This study identified the main factors limiting pea yields at 9 sites in 7 European countries (Spain, France, England, Belgium, Switzerland, Germany and Denmark) and classify the two study years—1998 and 1999 among the serie of years studied. The determination of the risks of occurrence of the main limiting factors of peas and the identification of the most favourable regions for pea growing or conversely those where the yield potential will be limited by various factors enables us to define the zones where peas, or indeed other legumes (ongoing work), are well suited in the different European countries taking part in the trials and to propose changes to cultural practices for the sites subject to limitation. This work has also allowed a validation of the pea crop model over a very large geographical area. Lastly, this study has, among other things, served to analyse the interactions between genotype and environment (Courty, 1999; cf. next part on genotype x environment interaction for yield and protein).

Conclusion

The development model is able to reconstitute retrospectively the main stages of pea growth. The climatic variables calculated for these key periods of growth are able to identify the limiting factors due to the weather and to explain why certain components were affected, compared with the potential with no constraints. Other variables characterising the crop (NNI at BF, harvest index etc.) can also help diagnose problems with the crop. These variables may be used to rank the limiting factors on the scale of a small region or to compare regions with one another on the national scale or to compare years. These variables turn out to be applicable on the European scale.

In view of these results, once the problem or problems have been identified, it is possible in certain cases to apply solutions. For example, early sowing or an application of water to a crop may help remedy a water stress habitually experienced in a region. Growing winter peas can also be considered to avoid at least part of the water and heat stress towards the end of growth in the regions most affected by high temperatures and water stress, since their developmental cycle is earlier than that of spring peas.

Methodology file: measurements of seed profiles and vertical structure

Isabelle Chaillet

Number of seeds per node (seed profile)

Pick up around 50 stems (3x15 to 20 consecutive stems).

On each stem of the sample, count the number of seeds on each node. The first node is the one that was the first to flower, from the base of the plant.

Make the mean of the number of seeds for each node.

A quicker method consist to gather together for each flowering node the pods of the different stems and then to count the number of seeds on each node. The mean number of seeds per node corresponds to the number of seeds per node divided by the number of stems of the sample.

Remark: Whatever the method, when a flowering node on a stem does not bear pod, this node is identify such as a fruit-bearing node without seeds.

The seed profile can be represented by the graph of the relationship between the number of flowering nodes and the number of seeds per node.

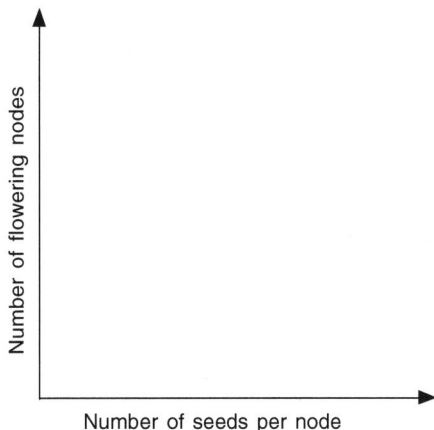

Number of seeds per node

Measurement of the vertical structure (or simplified profile)

Pick up around 50 stems (3x15 to 20 consecutive stems). On each stem, considering that 1 correspond to the first flowering node, one can note:

— the number of the first reproductive node (basal nodes may have aborted);
— the number of the last reproductive node;
— the number of the last flowering node (= number of flowering nodes). It is however often difficult to note the last flowering node;
— the number of reproductive nodes (bearing pods).

For each variable, the mean must be calculated on all the stems picked up.

10

Genotype x environment interaction for yield and protein concentration

Véronique Biarnès

Pea yields vary greatly between years and areas (Cf. previous chapter on diagnostic). There are also great differences in yield between varieties in a given location. These differences may vary, depending on the environment. In extreme situations, the ranking of varieties differ from one location to another: this is genotype x environment interaction. To explain the differences in behaviour of pea varieties in different locations, the knowledge of the crop's physiology in the first part of this book can be useful.

The presence of genotype x environment interaction for yield and protein concentration can complicate the evaluation of varietal performance for these two criteria. To investigate this, different studies were carried out on the analysis of interactions between variety and environment factors for yield (Dumoulin (1994), Courty (1999), Ragonnaud (2001)) and for protein concentration (Ragonnaud, 2001). These different studies show that ranking may vary considerably with the environment for yield but less so for protein concentration. They also identified genotypic characteristics that lead to differences in behaviour between varieties for these two criteria depending on the environment, which can help the breeder with varietal selection.

Importance of genotype x environment interactions for yield

Genotype x environment interactions

As for many other species, the results of the analysis of variance for yield in pea show that effects linked to the environment are very important: the environmental effect accounts for 60–90 % of the total variance, depending on the environment (Dumoulin (1994), Courty (1999)). In a study of two data sets of GEVES in 1999 and 2000 (Ragonnaud, 2001), the environmental effect was the main source of variability for yield (tables 3.4 and 3.5).

The genotype x environment interaction effect for yield is always significant and fairly considerable. It represents 5–20 % of the total variance (Dumoulin (1994), Courty (199)) and is similar to (Dumoulin, 1994) or from 2 (Courty, 1999) to 3 or 4 times (tables above) higher than the genotype effect. The ranking of varieties varies greatly among environments. Some varieties have opposite behaviour in contrasting environments. There can be a complete reversal of ranking in specific environments: On the other hand, intermediate varieties are the same for all the sites (Courty, 1999).

Table 3.4. Importance of variance components linked to genotype, environment and genotype x environment effects (1st data set : 2d year CTPS in 1999 and 1st year in ITCF-UNIP network in 2000).

Source of variability	Ratio to the variance linked to the genotype effect	
	Yield	Protein concentration
Genotype	1	1
Environment	22.4	1.4
Genotype x environment interaction	3.9	0.5

Table 3.5. Importance of variance components linked to genotype, environment and genotype x environment effects.

(2d data set : 1st year CTPS in 1999 and 2d year CTPS in 2000)

Source of variability	Ratio to the variance linked to the genotype effect	
	Yield	Protein concentration
Genotype	1	1
Environment	31.2	4.4
Genotype x environment interaction	3.3	1

Genotype x location and genotype x year interactions

The environment effect, which can be broken down into a year effect and a location effect shows that, for yield, genotype x year and genotype x location effects were significant at nearly all the studied situations (Ragonnaud, 2001). The ranking of genotypes is thus quite variable depending on years and locations.

It is therefore justified to evaluate genotypes (tests for agronomic and technological performance) when registering tests of varieties for yield over 2 years and on a quite large number of locations. However genotype x year has been found to be non-significant in previous studies (Dumoulin, 1994; Courty, 1999) that were carried out in only 2 years (1991 and 1992; 1998 and 1999 respectively) which did not vary greatly in terms of climate. However, Ragonnaud (2001) studied combinations of 3 or 4 years including 1996, which was very dry, and which certainly induced differences in varietal ranking compared to more "usual" years.

Characteristics that can induce particular genotype behaviour

To better understand differences in varietal ranking among environments, the statistical method of factorial regression may be used. Characteristics measured on genotypes in different environments (called genotypic covariates) and variables that describe environmental conditions (called environmental covariates) are introduced into the analysis of variance. Factorial regression allows these variables to be arranged in a hierarchy, depending on their influence on the ranking of varieties.

In the 3 examples where factorial regression has been used to explain genotype x environment interaction for field pea yield (table 3.6), a high level of explanation has been reached (from 50 % to 70 %) with relatively few covariates (3 to 5). This is very satisfactory (Brancourt-Hulmel et al., 1997).

Among genotypic covariates, earliness of flowering has appeared determinant in genotype x environment interaction, through the 1st flowering node (N1F) (Dumoulin, 1994) or date of beginning of flowering (BF) (Ragonnaud, 2001). These variables determine the beginning of the period of seed set on the plant (cf. p. 26–28 and p. 105; fig. 1.10).

Different variables related to the last of flowering time have also appeared essential in the explanation of interactions: the number of reproductive nodes (NR) (Dumoulin, 1994; Courty, 1999) or covariates closely related to the number of reproductive nodes elaborated, such as plant height at the end of flowering and date of the end of flowering (Ragonnaud, 2001). Note that NR fixes the end of the seed setting period, that is an important phase of the cycle.

Table 3.6. Analysis of the genotype x environment interactions for pea yield: results obtained with the factorial regression method

Authors, year	Dumoulin, 1994	Courty, 1999	Ragonnaud, 2001
Genotypic Covariate	N1F NR	Height at maturity TSW NR Lodging	Date of BF Height at EF Date of EF Earliness to maturity
Environmental Covariate	WB during flowering	WB before flowering	—
% of the GxE interaction depending Sum of Square explained by the model	64 %	> 68 %	50 % à 70 % on the data set

N1F : First flowering node
NR : Number of reproductive nodes
WB : Water balance
TSW : Thousand Seed Weight
BF : Beginning of flowering
EF : End of flowering

For environmental covariates, water balance just before (Courty, 1999) and during flowering time (Dumoulin, 1994) seems to be important in the ranking of genotypes for yield. No covariate related to the water balance emerged from the study conducted by Ragonnaud in 2001 because water conditions were relatively favourable for plant development in all the environments. Some sites were irrigated if necessary.

All of these results suggest that a substantial part of genotype x environment interaction for pea yield is due to differences in genotype behaviour when grown in environments with various water conditions, depending on their earliness of flowering (estimated by N1F or BF) and the end of seed setting (determined by NR or related variables: height at EF or date of EF).

However, harvest conditions are also very important to determine the ranking of genotypes for yield, especially in differing environments. Thus Ragonnaud (2001) showed earliness of maturity to be an important criterion. This is probably linked to problems that were encountered at harvest in France in 2000 (strong rains in July that delayed the harvest). At some sites, indeed, a late harvest might have harmed the earlier genotypes more than the later ones, because of over-maturity.

Furthermore, plant height at physiological maturity and sensitivity to lodging have appeared dominant in the European network (Courty, 1999), where variability of the climate was important. These two criterions explain yield losses (direct or linked to disease development) caused by lodged plants when harvest conditions are difficult (very dry weather and stony soil or conversely, very wet conditions). Thus, in 1999, at a Spanish site, harvested when the plants were very dry (seeds at 10–11 % water concentration) and in a relatively stony soil, the three most productive varieties were tall and had a large number of reproductive nodes and good resistance to lodging. The yield losses of mechanical origin were less for those varieties whose plants were still standing than for those completely lodged. This would explain their higher yield. At a German site, frequent heavy rain and violent winds at the end of growth enhanced lodging of the varieties in 1999. Two varieties remained standing and suffered limited yield losses (principally due to disease development), in contrast to varieties more sensitive to lodging.

Weak genotype x environment interactions for protein concentration

Genotype x environment interactions

Tables 3.4 and 3.5 (Ragonnaud 2001) show that the environment effect represents the main source of variation for protein concentration. The genotype x environment interaction may be of similar magnitude to, or as little as half of the variety effect. The ranking of genotypes for protein concentration seems to be very stable from one site to another. In nearly all the environments we find the same varieties at the top. For intermediate varieties, some inversions of ranking exist but differences between varieties are often around 0.5 % protein concentration, close to the level of precision of measurement and not statistically significant.

The same conclusions can be drawn from the data measured in the European network (Courty, 1999), where climate patterns differed more : ranking of varieties for protein concentration change little from one site to another.

Genotype x location and genotype x year interactions

For protein concentration, varietal rankings generally differ little from one year to another, except for dry years. However, in the study made by Ragonnaud (2001), genotype x year interaction was significant for two sets of data in only two years including 1996, which was marked by a drought stress that began very early. In these situations, where intense and prolonged drought stress can serious impair nitrogen fixation (Cf. part II.1.3.), genotype rankings are completely changed compared to a normal year when nitrogen nutrition can take place in good conditions. It is therefore necessary to judge this criterion over two years. The genotype x location interaction has been found to be significant in half of the studied cases, so it is not negligible.

It appears necessary therefore to test varieties at a number of locations, sufficient to take account of this variability. Compared to the trials generally carried out by CTPS, a further study showed that it was possible to reduce the number of sites used each year to evaluate genotypes by half (6 sites instead of 12).

Characteristics that can induce particular genotype behaviour

For protein concentration, it appears that genotype behaviour depends on seed weight and on the weather during the seed filling period (Ragonnaud, 2001). Small-seeded varieties have different ways to elaborate protein, depending on the environmental conditions, than those with large seeds.

However, the use of a sensitivity score to pea common mosaic virus (PCMV), evaluated by the GEVES in controlled conditions in tests of DHS (Distinction, Homogeneity, Stability), was also effective in explaining genotype x environment interaction for protein concentration in a data set where numerous sites experienced attack by aphids, which can transmit the virus. When there was infection, sensitive varieties had a lower protein concentration than resistant varieties.

Can varieties be stable for both yield and protein concentration?

In order to see if there are varieties with both high and stable yield and protein concentration, a study on different statistical parameters of stability for these two criteria was undertaken. A variety which is stable for these two variables of economic interest may guarantee the farmer reliable behaviour under most environmental conditions.

The ecovalence is a statistical parameter that evaluates stability for a given variable. It is the contribution of a variety to the sum of square of differences linked to the genotype x environment interaction. This parameter was calculated on two data sets (Biarnès et al., 2002). The combination of ecovalence (as a stability criterion) and the mean level of performance, simultaneously for yield and protein concentration, identifies two varieties, with very high yield and protein concentration in all environments.

The use of the ecovalence parameter allows varieties to be identified which are both productive and stable, and so reassuring for the farmer.

The lack of a relationship between ecovalence and mean varietal performances shows that performance and stability are not incompatible.

However, later results on these two varieties, collected in 2001, show that they can be less productive and stable than we would expect. That year was characterized by very late sowings and a severe drought during the flowering period, which might severely reduce yield and protein concentration elaboration in these two varieties. In the study, the years concerned (1999 and 2000) were much more favourable as regards climate. This underlines the limitations in the use of statistical parameters: they depend greatly on the studied data. It appears necessary to have many years of data to have representative sample and to find more reliable results.

Conclusion

Genotype x environment interactions are larger for yield than for protein concentration. For yield, ranking of varieties varies greatly, depending on locations and years. It is therefore necessary to evaluate varieties at numerous sites (at least 6) and over at least two years. Genotype rankings for protein concentration, on the other hand, are very stable over different locations and years, except for dry years.

The ecovalence, which indicates the contribution of each variety to the whole genotype x environment interaction, allows varieties which are stable both for yield and protein concentration to be identified. However for reliable results, many years of data are needed.

Thus, earliness of flowering and maturity, date of beginning and end of flowering, number of reproductive nodes and height at the end of flowering should largely determine the different behaviour of genotypes among sites for yield. Plant height at maturity and resistance to lodging should also play an important role in areas where the climatic conditions at the end of the cycle may be unfavourable (very dry or very wet weather). For protein concentration, seed weight and resistance to Pea Common Mosaic Virus (PCMV) appear to be mainly responsible of different behaviour among varieties.

The breeder can modify these different characteristics to better adapt pea varieties to a wide range of environments and to guarantee stable yield and protein concentration. He can also influence the characteristics that determine the beginning (N1F, BF) and the end (NR, EF, plant height at EF) of the seed setting period (the most sensitive period in the cycle) and try to arrange the reproductive development of varieties to match weather conditions, especially rainfall, meanwhile attaching importance to the plant height at maturity and to standing ability. The integration of resistance to PCMV is also valuable to guarantee a high level of protein concentration even when there is an attack of aphids, which can transmit the virus.

An understanding of the reproductive development model (cf. fig. 1.10) and of the genetic and environmental determination of characteristics involved in flowering (cf. part I.2.) may help the breeder to produce more stable varieties.

Prospects for legume crops in France and Europe[2]

Nathalie Munier-Jolain, Benoît Carrouée

This book is an overview of the knowledge coming from research conducted mainly in France during the eighties. This major investment in research for understanding grain legume crops has met the requirements of the EEC which aims to increase legume crops in Europe in order to reduce the shortage of plant proteins. Over the course of twenty years, grain legume crops, which were seldom cultivated in Europe at the beginning of the eighties, expanded to cover up to 630 000 ha in France in 1997 and 1998, but then dropped to approximately 350 000 ha in 2000. This expansion of pea growing has been accompanied by a yield increase of 4 % per year, partly due to genetic progress, but also to the improvement of production techniques allowed by research in agronomy, ecophysiology, pathology, etc. However 20 years later the shortage of plant proteins remains significant. Only 26 % of the requirements of plant proteins for animal feeding are met by European products, whereas 60 % are satisfied with imported soybean.

During this period the environmental impacts of agricultural activities have become more and more crucial for agricultural specialists and decision makers, because the analysis of those impacts guided the environmental regulations and calls for the introduction of incentives.

Although the requirements for protein-rich raw materials are huge and the environmental considerations are becoming important, only 5 % of the cultivated areas in Europe are sown with legume crops (soybean, pea, lupin, faba bean, lucerne, etc.), whereas these crops represent 20 to 30 % of the cultivated area in America and Australia. The research conducted over many years can help to support the major advantages of legume crops in sustainable cropping systems.

---

2. This conclusion is a synthseis of an environmental review on pea published in Cahoers EAgriculture (Munier-Jolain and Carrouée, 2003).

Quality products for animal and human nutrition

Peas are easily stored and processed by both the industry and animal breeders. Pea seeds constitute a valuable material for animal feeding, largely because they have high starch and digestible protein concentrations, with a very high lysine/protein ratio. Moreover peas, by virtue of their traceability (freedom from GMO (genetic modified organisms) and their composition, are greatly valued in "labeled" or organic meat production programs.

Peas are a crop with low requirements

The water requirements of peas are low because of their short growing period. Irrigation in northern regions is either unnecessary or else does not compete with other spring crops whose water requirements generally occur later. Thus water requirements of peas are not only low, but they are also fit in well with the water management of the farm.

Having the ability to symbiotically fix atmospheric dinitrogen (N_2), peas produce high-protein seeds without any nitrogen fertilization. Supplying N does not increase either seed yield or N concentration. Fertilizer manufacture represents the largest consumption of fossil energy by agriculture (essentially natural gas for fertilizers manufactured in Europe); e.g. nitrogen fertilisation represents 60 % of the energy cost for a wheat crop receiving 180 kgN ha^{-1}. This lower consumption of fossil energy for legume crops compared to other crops leads to a similar decrease in CO_2 emissions, and concomitantly of N_2O. Because of their low requirements for pesticides and fertilizers, peas have lower energy costs than wheat. It also reduces the emissions of N_2O, NO_x and CH_4, which, in addition to the CO_2 saving, significantly reduces the greenhouse effect due to agricultural activities.

Consequently the ability to symbiotically fix atmospheric dinitrogen constitutes an advantage as regards climate change. Although data on gaseous emissions are imprecise, a proportion of 25 % of legume crops in European crop rotations could reduce greenhouse gas emissions by approximately 1 %. Such a decrease of greenhouse gas emissions is not negligible compared with the target range of reduction of greenhouse gas emissions fixed by the Kyoto protocol.

A solution for the diversification of crop rotation

Chemical management of weeds is costly for farmers and highly risky for the environment. In France it has led to the selection of a specific weed flora in winter crops and of herbicide-resistant weeds in cereal cropping systems. In such systems pea presents a way to diversify the crop rotation. It also allows farmers to vary (i) the nature of the herbicides being applied to the crop rotation and (ii) sowing dates. The latter represent a non-chemical means to reduce weed infestation and to diversify the weed flora. However the benefits for weed control of the different types of pea crop (winter or spring sown) depend on the timing of sowing compared to the emergence date of the weeds generally found in cereals.

Similarly, the introduction of a pea crop into cereal rotations reduces the influence of parasites of other crops (root or shoot diseases), because there are few pests which attack pea crops as well as cereals and oilseed crops (except Sclerotinia, which also attacks oilseed rape and sunflower).

Reducing the risks of nitrate pollution

Because of its autonomous nitrogen nutrition, pea avoids environmental risks due to both the manufacture, transport, and use of chemical nitrogen fertilizers. When availability of soil mineral nitrogen is low, pea nitrogen nutrition relies mainly on symbiotic N_2 fixation. As soil mineral nitrogen availability increases, uptake and assimilation of nitrates efficiently complement symbiotic fixation to give optimal nitrogen nutrition. Thus nitrogen nutrition of pea cannot itself cause nitrate leaching.

The difference between the amount of nitrogen exported to the seeds and that accumulated by the crop (in root and aerial parts, plus exudates) is in most of the cases as low for pea as is observed for wheat. This balance is even negative when straw is exported, because pea retrieves more nitrate from the soil than it returns to the soil after harvest. However high levels of residual soil nitrogen are sometime found after a pea crop. At the beginning of winter, levels of residual soil nitrate can be approximately 30 units higher than for cereals, but are roughly the same as those after oil seed rape, although nitrogen nutrition and overall N balance are very different for the two crops. These may result from either (i) a low nitrogen uptake due to the shallow pea root system or more likely (ii) early maturity, which prevents the crop from retrieving all the nitrogen arising from the summer mineralization of organic matter.

Even when nitrogen mineralization is high during the summer, risks of nitrate leaching during the following winter are still no greater for a pea crop than for many other crops. However when the risks of nitrogen leaching are high and because of the early maturity of pea a ICNT (intermediate crop "nitrate-trap") may be sown before wheat.

Because mineralization of pea residues is limited in autumn and occurs essentially in spring, the pea crop constitutes a good preceding crop for wheat; the nitrogen supplied by mineralization in spring allows the nitrogen fertilization for the following crop to be reduced by about 30 units.

Consequently the problem of nitrate pollution after a pea crop could be easily solved with (i) better management of nitrogen fertilization for the preceding and following crops, (ii) the development of new practices such as nitrate-trapping catch crops, and (iii) the improvement of the root system both by breeding and by sowing without any soil structure degradation.

Concluding remarks

Because of their paramount and original role in the management of environmentally-friendly cropping systems, peas should attract more interest when sustainable agriculture is promoted. Indeed peas have considerable intrinsic quality characteristics which suit the needs of the "labeled" animal feed industry, and have low requirements for water, nutrients and energy. Besides, in order to establish innovative forms of agriculture, diversifying crop successions is a prerequisite in order to reduce pesticide (particularly herbicide) use on rotations, without compromising the sustainability of cropping systems. This implies the mastering of nitrogen budgets in crop successions, especially during the winter periods that precede and follow peas.

The pea crop has major advantages in sustainable farming systems: (i) its nutritional characteristics are well suited to quality meat husbandry; (ii) pea has low requirements for water, chemicals and fossil energy; (iii) its ability to symbiotically fix atmospheric nitrogen precludes the need for N-fertilizer; (iv) its use may help in diversifying crop rotation, allowing a lower input of pesticides to the rotation.

Nevertheless, the development of the pea crop in crop rotations, requires joint efforts of researchers, breeders and all actors of the agricultural technical development to increase part of protein crops in crop rotations and also to conquer new cropping areas.

Bibliography

Aitken Y., 1978. Flower initiation in relation to maturity in crop plants. IV. Sowing time and maturity type in pea (*Pisum sativum* L.) in Australia. *Aust. J. Agric. Res.*, 29, 983–1001.

Alamie J., 2002. Caractérisation d'un réseau multilocal de parcelles infestées par Aphanomyces euteiches, champignon phytopathogène du pois (*Pisum sativum* L.): étude des interactions génotype x milieu et perspectives pour la sélection. Mémoire de fin d'études de DESS Gestion des Agroressources. Université de Reims Champagne Ardennes, 30 pp.

Amarger N., 1991. Fixation de l'azote moléculaire par les associations légumineuses-bactéries en symbiose: aspects génétiques et agronomiques. *C.R. Acad. Agric.*, 71, 143–152.

Amarger N., Mariotti A., Mariotti F., Dürr J.C., Bourguignon C. and Lagacherie B., 1979. Estimate of symbiotically fixed nitrogen in field grown soybeans using variations in ^{15}N natural abundance. *Plant Soil*, 52, 269–280.

Armstrong E.L. and Pate J.S., 1994. The field pea in south west Australia I. Patterns of growth, biomass production and photosynthesis performance in genotypes of contrasting morphology. *Aust. J. Agric. Res.*, 45, 1347–1362.

Armstrong E.L., Pate J.S. and Tennant D., 1994. The field pea in south western Australia. Patterns of water use and root growth in genotypes of contrasting morphology and growth habit. *Aust. J. Plant Physiol.*, 21, 517–532.

Arumingtyas E.L., Floyd R.S., Gregory M.J. and Murfet I.C., 1992. Branching in *Pisum* inheritance and allelism tests with 17 ramosus mutants. *Pisum genetics*, 24, 17–31.

Arumingtyas E.L. and Murfet I.C., 1994. Flowering in *Pisum*: a further gene controlling response to photoperiod. *J. Hered.*, 85, 12–17.

ARVALIS Institut du Végétal 2003. Quoi de neuf? Pois, féveroles & lupins 2003. 89 p.

Atkins, G.A., Kuo J. and Pate J.S., 1977. Photosynthetic pod wall of pea (*Pisum sativum* L.). *Plant Physiol.*, 60, 779–786.

Atkins C.A., Pate J.S., Griffiths G.J. and White S.T., 1980. Economy of carbon and nitrogen in nodulted and non-nodulated (NO$_3^-$ cowpea [*Vigna unguiculata* (L.) Walp.]. *Plant Physiol.*, 66, 978–983.

Atkins C.A., Pate J.S., Sanford P.J., Dakora F.D. and Matthews I., 1989. I. Nitrogen nutrition of nodules in relation to 'N-hunger' in cowpea (*Vigna unguiculata* L.Walp). *Plant Physiol.*, 90 (4), 1644–1649.

Bakry M.O., El-Magd A.M.M. and Shaheen A.M., 1984. Response of growth and yield of pea (*Pisum sativum* L.) to plant population and NPK fertilization. *Egypt. J. Hort.*, 11, 151–161.

Bandyopadhyay, A.K., Veena Jain and Nainawatee H.S., 1996. Nitrate alters the flavonoid profile and nodulation in pea (*Pisum sativum* L.). *Biol. Fert. Soils*, 21 (3), 189–192.

Barber H.N., 1959. Physiological genetics of *Pisum*. II. The genetics of photoperiodism and vernalisation. *Heredity*, 13, 33–60.

Bassanezi R.B., Amorim L., Bergamin F.A., Hau B. and Berger R.D., 2001. Accounting for photosynthetic efficiency of bean leaves with rust, angular leaf spot and anthracnose to assess crop damage. *Plant Pathol.*, 50, 443–452.

Bastiaans L., 1991. Ratio between virtual and visual lesion size as a mesure to describe reduction in leaf photosynthesis of rice due to leaf blast. *Am. Phytopathol. Soc.*, 81, 611–615.

Bastiaans L., 1993. Effects of leaf blast on photosynthesis of rice. I. Leaf photosynthesis. Neith. *J. Plant Pathol.*, 99, 197–203.

Bauer W.D., 1981. Infection of legumes by *rhizobia*. *Ann. Rev. Plant Physiol.*, 73, 407–449.

Béasse C., 1998. Impact de l'anthracnose sur la croissance et le développement du Pois protéagineux en condition de peuplement. Thèse, Université de Rennes I, 100 p.

Béasse C., Ney B. and Tivoli B., 1999. Effects of pod infection by *Mycosphaerella pinodes* on yield components of pea (*Pisum sativum*). *Ann. Appl. Biol.*, 135, 259–367.

Béasse C., Ney B. and Tivoli B., 2000. A simple model of pea (*Pisum sativum*) growth affected by Mycosphaerella pinodes. *Plant Pathol.*, 49, 187–200.

Beck D.P., Wery J., Saxena M.C. and Ayadi A., 1991. Dinitrogen fixation and nitrogen balance in cool-season food legumes. *Agron. J.*, 83, 334–341.

Bélanger G., Gastal F. and Lemaire G., 1992. Crop ecology, production and management: growth analysis of tall Fescue saward fertilized with different rate of nitrogen. *Crop Sci.*, 32, 1371–1376.

Bergersen F.J., Turner G.L. Peoples M.B., Gault R.R., Mothorpe L.J. and Brockwell J., 1992. Nitrogen fixation during vegetative and reproductive growth of irrigated soybean in the field: application of ^{15}N methods. *Aust. J. Agric. Res.*, 43, 145–153.

Berry G.J. and Aitken Y., 1979. Effect of photoperiod and temperature on flowering in pea (*Pisum sativum* L.). *Aust. J. Plant Physiol.*, 6, 573–587.

Bethlenfalways G.J., Abu-Shakra S.S., Fishbeck K. and Philips D.A.,1978. The effect of source-ink manipulations on nitrogen fixation in peas. *Plant Physiol.*, 43, 31–34.

Bhuvaneswari T.V., Turgeon B.G. and Bauer W.D., 1980. Early events in the infection of soybean (Glycine max L. Merr.) by Rhizobium Japonicum. I. Localization of infectible root cells. *Plant Physiol.*, 66, 1027–1031.

Biarnès V., 2003. D'exceptionnels dégâts de gel pour un hiver atypique. *Persp. Agric.*, 293, 22–25.

Biarnès V., Faloya V. and Chambenoit C., 2000. Teneur en protéines des graines de protéagineux : les effets du milieu et de la variété. *Persp. Agric.*, 261, 2–4.

Biarnès V., Roullet G., Baril C. and Ragonnaud G., 2002. Rendement et teneur en protéines du pois protéagineux: mieux comprendre les interactions variétés x milieux. *Persp. Agric.*, 281, 12–17.

Biarnès-Dumoulin V., Gaillard B., Mangin M., Ney B. and Wéry J., 1998. Modelling field pea yield: a tool for crop diagnosis and prospects. 3rd European Conference on Grain Legumes, Valladolid, Spain, 322–323.

Bindi M., Sinclair T.R. and Harrison J., 1999. Analysis of seed growth by linear increase in harvest index. *Crop Sci.*, 39, 486–493.

Blight M.M., Pickett J.A., Smith M.C. and Wadhams L.J., 1984. An aggregation pheromone of *Sitona lineatus*. *Naturwissenschaften*, 71, 480.

Boiffin J., Caneill J., Meynard J.M. and Sebillote M., 1981. Elaboration du rendement et fertilisation azotée du blé d'hiver en Champagne crayeuse. I-Protocole et méthode d'étude d'un problème technique régional. *Agronomie*, 7, 549–58.

Bonhomme R., 2000. Beware of comparing RUE values calculated from PAR vs solar radiation or absorbed vs. intercepted radiation. *Field Crops Res.*, 68, 247–252.

Bonhomme R., 2000. Review: Bases and limits to using "degree.day" units. *Eur. J. Agron.*, 13, 1–10.

Bouchard C., 1997. Influence de dynamiques variables de nutrition azotée sur la croissance et l'élaboration du rendement d'une culture de blé tendre d'hiver (*Triticum aestivum*). Mémoire d'ingénieur ITIA, 67 p.

Bourion V., Duparque M., Lejeune-Hénaut I. and Munier-Jolain N., 2002. Criteria for selecting productive and stable pea cultivars. *Euphytica*, 126, 391–399.

Bourion V., Lejeune-Hénaut I., Munier-Jolain N.G. and Salon C., 2003. Cold acclimation of winter and spring peas: carbon partitioning as affected by light intensity. *Eur. J. Agron.*, 19, 535–548.

Bournoville R., Badenhausser I. and Cantot P., 1990. Appréciation des risques liés aux ravageurs principaux des cultures de pois protéagineux.

In ANPP (ed), 2ème Conférence Internationale sur les ravageurs en agriculture, Versailles, France 3–6 décembre 1990, vol III, 1075–1086.

Boyce P.J. and Volenec J.J., 1992. Taproot carbohydrate concentrations and stress tolerance of contrasting alfalfa genotypes. *Crop Sci.*, 32, 757–761.

Brancourt-Hulmel M., Biarnès-Dumoulin V. and Denis J.B., 1997. Points de repère dans l'analyse de la stabilité et de l'interaction génotype x milieu en amélioration des plantes. *Agronomie*, 17, 219 246.

Brancourt-Hulmel M., Lecomte C. and Meynard J.M., 1999. A Diagnosis of Yield-Limiting Factors on Probe Genotypes for Characterizing Environments in Winter Wheat Trials. *Crop Sci.* 39, 1798–1808.

Brand J.M., Armstrong R. and Antonoff G., 2003. The effect of sowing date and rate on seed coat discouloration due to frost in field peas in southern Mallee of Victoria. *In:* Solutions for a better environment, 11th Australian Agronomy Conference. Australian Society of Agronomy Eds, Geelong, 4 p.

Bretag T.W., 1991. Epidemiology and control of Ascochyta blight of field peas. Ph.D. Thesis. School of Agricultural and Department of Botany, School of Biological Sciences. La Trobe University, Bundoora, Victoria, 3083, Australia, 297 p.

Brewin N.J., 1991. Development of the legume root nodule. *Ann. Rev. Cell Biol.*, 7, 191–226.

Brun D., 2002. Développement du pois d'hiver en zone Nord. DAA Systèmes de Production Agricole, ENSAIA, Nancy, 33 p.

Burstin J., Marget P., Huart M., Aubert G., Munier-Jolain N., Duchêne C., Desprez B. and Duc G., 2003. Genetic dissection of the variability of pea seed nutritional value. QTLs of seed protein content in pea. 64 p. Génoplante séminaire. Poitiers.

Butler J.H. and Ladd J.N., 1985. Symbiotically-fixed and soil-derived nitrogen in legumes grown in pots in soils with different amounts of available nitrate. *Soil Biol. Biochem.*, 17 (1), 47–55.

Caetano-Anolles G. and Gresshoff P.M., 1990. Early induction feedback regulatory responses governing nodulation in soybean. *Plant Sci.*, 71, 69–81.

Caetano-Anollès G. and Gresshof P.M., 1991. Plant genetic control of nodulation. *Ann. Rev. Microbiol.*, 45, 345–382.

Cantot P., 1986. Quantification des populations de *Sitona lineatus* L. et de leurs attaques sur pois protéagineux *(Pisum sativum L.). Agronomie*, 6 (5), 481–486.

Carberry P.S., Muchow R.C. and Hammer G.L., 1993. Modelling genotypic and environmental control of leaf area dynamics in grain sorghum. II-Individual leaf level. *Field Crops Res.*, 33, 329–351.

Carpenter J.F., Hand S.C., Crowe L.M. and Crowe J.H., 1986. Cryoprotection of phosphofructokinase with organic solutes: characterization of enhanced protection in the presence of divalent cations. *Arch. Biochem. Biophys.*, 250, 505–512.

Carrouée B. and Duchêne E., 1993. Comment maîtriser de telles variations? Teneur en protéines du pois. *Persp. Agric.*, 183, 75–81.

Carrouée B. and Gatel P., 1995. *In*: Peas, Utilisation in animal feeding. UNIP ITCF.

Carrouée B. and LeSouder C., 1992. Prévision du risqué de fuite en azote derrière une culture de pois. *Persp. Agric.*, 183, 75–81.

Casal J.J., Sanchez R.A. and Deregibus V.A., 1986. The effect of plant density on tillering: the involvement of R/FR ratio and proportion of radiation intercepted per plant. *Env. Exp. Bot.*, 26, 365–371.

Castonguay Y., Nadeau P., Lechasseur P. and Chouinard L., 1995. Differential accumulation of carbohydrates in alfalfa cultivars of contrasting winterhardiness. *Crop Sci.*, 35, 509–516.

Catroux G., 1991. Normes de qualité pour inoculums et contrôles en France. Consultations d'experts sur la production et le contrôle de qualité des inoculums pour légumineuses, Rome.

Choudhury B.J., 2000. A sensitivity analysis of the radiation use efficieincy for gross photosynthesis and net carbon accumulation by wheat. *Agric. Forest Meteorol.*, 101: 217–234.

Cline M.G., 1991. Apical Dominance. *Bot. rev.*, 57, 318–359.

Cline M.G., 1994. The role of hormones in apical dominance. New approaches to an old problem in plant development. *Physiol. Plant.*, 90, 230–237.

Colnenne C., Meynard J.M., Reau R., Justes E. and Merrien A., 1998. Determination of a Critical Nitrogen Dilution Curve for Winter Oilseed Rape. *Ann. Bot.*, 81, 311–317.

Colnenne C., Meynard J.M, Roche R. and Reau R., 2002. Effects of nitrogen deficiencies on autumnal growth of oilseed rape. *Eur. J. Agron.*, 17(1), 11–28.

Combaud S., 1996. Croissance foliaire d'un couvert de pois (*Pisum sativum* L.) soumis à un déficit hydrique : Conséquences sur le rendement en graines et la consommation en eau. Thèse de doctorat en Sciences Agronomiques de l'ENSA M. 83 p.

Corre G., Crozat Y. and Aveline A., 2001. On-farm diagnosis of factors limiting pea yields in organic farming. 4th European Conference on grain legumes: towards the sustainable production of healthy food, feed and novel products. Cracow, Poland, 8–12 July 2001, 46–47.

Corre-Hellou G. and Crozat Y. 2004. N_2 fixation and N supply in organic pea (*Pisum sativum* L) cropping systems as affected by weeds and peaweevil (*Sitona lineatus L.*). *Eur. J. Agron.* (in press).

Cosgrove J., 1993. How do plant cell walls extend? *Plant Physiol.*,102,1–6.

Courty J., 1999. Analyse de la variabilité du rendement du pois protéigineux de printemps (*Pisum sativum* L.) dans un réseau européen. Etude des interactions génotype x milieu. Mémoire de fin d'études de l'ENSAIA Nancy, 41p. + annexes.

Crozat Y., 2000. Agrophysiologie et fonctionnement du peuplement cultivé: exemples de la fixation biologique de l'azote par les légumineuses et de la production de graines chez les espèces à croissance indéterminée. Mémoire d'habilitation à diriger des recherches, Université d'Angers, Tome 1 : Synthèse des travaux, 87p.

Crozat Y., Aveline A., Coste F., Gillet J.P. and Domenach A.M., 1994. Yield performance and seed production pattern of field-grown pea and soybean in relation to N nutrition. *Eur. J. Agron.*,3 (2), 135–144.

Crozat Y., Corre G. and Larmure A., 2000. Des pois plus rapides que les mauvaises herbes. *Persp. Agric.*, 253, 92–93.

Crozat, Y., Doré T. and Gillet J.P., 1992. Determination of standard relationships for an early prediction of the number of grains of the pea crop. 1ère Conférence européenne sur les protéagineux, Angers.

Crozat Y., Gillet J.P., Gille D. and Tricot F., 1991. Pois de printemps: effets de la structure du sol et de l'eau sur la nutrition azotée et le rendement. *Persp. Agric.*, 162, 53–58.

Crozat Y., Gillet J.P. and Tricot F., 1991. Effects of the soil compaction on root distribution N, P, K uptakes and growth of pea crop. Proc. of the 3rd International Symposium of the Society of root research, Vienna.

Crozat Y., Gillet J.P., Tricot F. and Domenach A.M., 1992. Nodulation, N_2 fixation and incorporation of combined N from soil into pea crop: effects of soil compaction and water regime. First European Conference on Grain Legumes, Angers (France), Paris: A.E.P.

Cutter E.G. and Chiu H.W., 1975. Differential responses of buds along the shoot to factors involved in apical dominance. *J. Exp. Bot.*, 26, 828–839.

Danyluk J., Perron A., Houde M., Limin A., Fowler B., Benhamou N. and Sarhan F., 1998. Accumulation of an Acidic Dehydrin in the Vicinity of the Plasma Membrane during Cold Acclimation of Wheat. *Plant Cell*, 10, 623–638.

Davies W.J. and Tardieu F., Trejo C.L., 1994. How do chemical signals work in droughted plants? *Plant Physiol.*, 104, 309–314.

Dazzo F.B. and Brill W.J., 1978. Regulation by fixed nitrogen of host-symbiont recognition in the Rhizobium-clover symbiosis. *Plant Physiol.*, 62 (1), 18–21

Demotes-Mainard S. and Jeuffroy M.H, Robin S., 1999. Spike dry matter and nitrogen accumulation before anthesis in wheat as affected by nitrogen fertilizer: relationship to kernels per spike. *Field Crops Res.*, 64, 249–259.

Den Boer B.G.W. and Murray J.A.H., 2000. Triggering the cell cycle in plants. *Trends Cell Biol.*, 10, 245–250.

Dereuddre J. and Gazeau C., 1992. Les végétaux et les très basses températures. In Les végétaux et le froid. D. Côme Eds, Hermann, Paris, 600 p.

Devienne-Barret F., Justes E., Machet J.M. and Mary B., 2000. Integrated control of nitrate uptake by crop growth rate and soil nitrate availability under field conditions. *Ann. Bot.*, 86: 995–1005.

Doré T., 1992. Analyse, par voie d'enquête, de la variabilité des rendements et des effets précédants du pois protéagineux de printemps (*Pisum sativum* L.). Thèse de Doctorat, INA-PG, Paris, 214 p.

Doré T., 1994. Influence sur l'évolution du nombre de ramifications et de tiges chez le pois. *Agrophysiologie du pois protéagineux, UNIP - ITCF - INRA*, mai, 145–154

Doré T. and Meynard J.M., 1995. On-farm analysis of attacks by the pea weevil (*Sitona lineatus* L Col, Cucurlionidae) and the resulting damage to pea (*Pisum sativum* L.) crops. *Journal of applied entomology*, 19, 49–54.

Doré T., Meynard J.M. and Sebillotte M., 1998. The role of grain number, nitrogen nutrition and stem number in limiting pea crop (*Pisum sativum* L.) yields under agricultural conditions. *Eur. J. Agron.*, 8, 29–37.

Drevon J.J. and Hartwig U.A., 1997. Phospohrus deficiency increases the argon-induced decline of nodule nitrogenase activity in soybean and alfalfa. *Planta*, 201, 463–469.

Du Montant E., 2001. Adaptation du pois d'hiver en Zone Nord. Mémoire de fin d'études DESS Protection et Valorisation du Végétal, Université de Pau et des Pays de l'Adour.

Dumont E., Fontaine F., Vuylsteker C., Sellier H., Bodèle S., Voedts N., Devaux R., Frise M., Avia K., Hilbert J.-L., Bahrman N., Hanocq E., Lejeune-Hénaut I. and Delbreil B., 2009. Association of sugar content QTL and PQL with physiological traits relevant to frost damage resistance in pea under field and controlled conditions. Theor. Appl. Genet., Published online: March 2009.

Dumoulin V., 1994. Etude de la variabilité génétique chez le pois protéagineux pour l'élaboration du rendement. Importance des interactions génotype x milieu. Thèse de doctorat INA-PG, 232 p.

Dumoulin V., Ney B. and Etévé G., 1994. Variability of pea seed and plant development in pea. *Crop Sci.*, 34, 992–998.

Duru M. and Lemaire G., 1997. Grasslands. *In: Diagnosis of the Nitrogen Status in Crops*, Lemaire G. (Ed), Springer, Berlin, 59–72.

Duthion C. and Pigeaire A., 1991. Seed lengths corresponding to final stage in seed abortion of three grain legumes. *Crop Sci.*, 31, 1579–1583.

Dwyer L.M. and Stewart D.W., 1986. Leaf area development in field grown maize. *Agron. J.*, 78, 334–343.

Dwyer L.M., Stewart D.W., Hamilton R.I. and Houwing L., 1992. Ear position and vertical distribution of leaf area in corn. *Agron. J.*, 84: 430–438.

Egli D.B., Ramseur E.L., Zhen-Wen Y. and Sullivan C.H., 1989. Source-sink alterations affect the number of cells in soybean cotyledons. *Crop Sci.*, 29, 732–735.

Ellis T.H.N. and Poyser S.J., 2002. An integrated and comparative view of pea genetic and cytogenetic maps. New Phytol., 153: 17–25.

Emery N.R.J., Longnecker N.E. and Atkins C.A., 1998. Branch development in *Lupinus angustifolius* L. II. Relationship with endogenous ABA, IAA and cytokinins in axillary and main stem buds. *J. Exp. Bot.*, 49, 555–562.

Engqvist L.G., 2001. Effects of infection of common root rot on protein content, cooking quality and other characters in pea varieties. *Nahrung*, 45 (6), 374–376.

Etévé G., 1985. Breeding for cold tolerance and winter hardiness in pea. In The Pea Crop. P.D. Hebblethwaite, M.C. Heath and T.C.K. Dawkins Eds, Butterworths, London, 486 p.

Evans L.T., 1994. Crop physiology: Prospects for the retrospective science. In 'Physiology and determination of crop yield'. Eds B.K.J., J.M. Bennett, T.R. Sinclair and Paulsen G.M. ASA, CSSA, SSSA, Madison, WI, USA. pp. 19–36.

Fahn A., 1990. 'Plant anatomy.' Pergamon Press: Oxford, England.

Feierabend J., Schaan C. and Hertwig B., 1992. Photoinactivation of catalase occurs under both high- and low-temperature stress conditions and accompanies photoinhibition of photosystem II. *Plant Physiol.*, 100, 1554–1561.

Field R.J. and Jackson D.I., 1974. Light effects on apical dominance. *Ann. Bot.*, 39, 369–374.

Flinn A.M., Atkins C.A. and Pate J.S., 1977. Significance of photosynthetic and respiratory exchanges in the carbon economy of the developing pea fruit. *Plant Physiol.*, 60, 412–418.

Flinn A.M. and Pate J.S., 1970. A quantitative study of carbon transfer from pod and subtending leaf to the ripening seeds of the field pea (*Pisum arvense* L.). *J. Exp. Bot.*, 21, 71–82.

FNAMS, 1997. Recherche des causes de fragilité des semences de pois protéagineux et tests d'itinéraires techniques pour la production de semences de qualité. Contrat de Branche 1994–1996. en partenariat avec l'INA-PG, la SNES, and l'ITCF. Rapport de synthèse 49 p.

Foucher F., Morin J., Courtiade J., Cadioux S., Ellis N., Banfield M.J., Rameau C., 2003. Determinate and late flowering are two terminal flower1/

centroradialis homologs that control two distinct phases of flowering initiation and development in pea. *Plant Cell*, 15: 1–14.

Fougereux J.A., 1994. Etude de l'influence de l'alimentation hydrique du porte-graine sur la qualité germinative des semences de pois protéaginuex (Pisum sativum L.). Thèse de doctorat de l'INA.PG. 160 p.

Fougereux J.A., 1999a. La fragilité des semences de pois : pourquoi ça casse ? *Bull. Semences*, 144, 16–18.

Fougereux J.A., 1999b. Qualité des semences de pois: elle se produit d'abord au champ. *Bull. Semences*, 150, 17–18.

Fougereux J.A., Doré T., Deneufbourg F. and Ladonne F., 1998a. Assessment of pea germination quality in farmer's field before mechanical harvest. Proceedings of the 3rd Conference on Grain Legumes, Valladolid, p 291.

Fougereux J.A., Doré T. and Ladonne F., 1998. A rapid method for determining seed set and the beginning of the seed filling stage in *Pisum sativum* L. 3rd European Conference on Grain Legumes. Valladolid. 471 p.

Fougereux J.A., Ducournau S., Mannino M., Ladonne F., Tyszka F. and Doré T., 1998b. Relationship between physical characteristics of the seed of Pisum sativum L. and its resistance to mechanical injury. Proceedings of the 3rd Conference on Grain Legumes–Valladolid, p 293.

Francisco P.B. and Akao S., 1993. Autoregulation and nitrate inhibition of nodule formation in Soybean cv. Enrei and modulation mutants, *J. Exp. Bot.*, 44 (260), 547–553.

Gallagher J.N. and Biscoe P.V. 1978. Radiation absorption, growth and yield of cereals. *J. Agric. Sci.*, 91, 47–60.

Garry G., 1996. Incidence de l'anthracnose à *Mycosphaerella pinodes* sur la synthèse des assimilats carbonés et azotés du pois protéagineux (*Pisum sativum* L) et leur transfert vers la graine: conséquences sur la formation et le remplissage des graines. Thèse, Université de Rennes I, 119 p.

Garry G., Jeuffroy M.H., Ney B. and Tivoli B., 1998b. Effects of ascochyta blight (*Mycosphaerella pinodes* Berk. and Blox.) on the decrease in photosynthesizing leaf area and the reduction of photosynthetic efficiency by green leaf area of dried-pea (*Pisum sativum* L.). *Plant Pathol.*, 47, 473–479.

Garry G., Jeuffroy M.H. and Tivoli B., 1998a. Effects of ascochyta blight (*Mycosphaerella pinodes* Berk. and Blox.) on biomass production, seed number and seed weight of dried pea (*Pisum sativum* L.) according to plant stage and disease intensity. *Ann. Appl. Biol.*, 132, 49–59.

Garry G., Tivoli B., Jeuffroy M.H. and Citharel J., 1996. Effects of ascochyta blight caused by *Mycosphaerella pinodes* on changes in carbohydrates and nitrogenous compounds in the leaf, the hull and the seed of dried-pea during seed filling. *Plant Pathol.*, 45, 769–777.

Georgieva K. and Licntenhaler H.K., 1999. Photosynthetic activity and acclimation ability of pea plants to low and high temperature treatment as studied by means of chlorophyll fluorescence. *J. Plant Physiol.*, 155, 416–423.

GNIS and FNAMS, 1992. Production de semences certifiées de pois protéagineux : récolte et stockage. Diaporama. 59 photos+commentaires.

Godin C., Costes E. and Sinoquet H., 1999. A method for describing plant architecture which integrates topology and geometry. *Ann. Bot.*, 84, 343–357.

Gouzonnat F., 1992. Test d'un modèle délaboration du nombre de graines d'un peuplement de pois protéagineux au champ sur quatre variétés. Mémoire Ingénieur ENITA.

Grandgirard D., Munier-Jolain N.G., Salon C. and Ney B., 2001. Nitrogen nutrition level and temperature effects on vegetative N remobilisation rate and distribution of canopy N during seed filling period in soybean (*Glycine max* L. Merr.). 4th European Conference on Grain Legumes, Cracovie. pp. 30–31.

Grandgirard D., 2002. Analyse et modélisation du déterminisme de la teneur en azote des graines chez le soja (*Glycine max* L. Merrill) : Relation entre la remobilisation d'azote vers les graines et l'élaboration du rendement et de la qualité. Thèse de doctorat de l'Université de Bourgogne. 159 p.

Greenwood D.J., Gerwitz A., Stone D.A. and Barnes A., 1982. Root development of vegetable crops. *Plant Soil*, 68, 75–96.

Guilioni L., 1997. Effet des hautes températures sur la croissance végétative et le développement des organes reproducteurs chez le pois (*Pisum sativum* L., cv. Messire). Conséquences sur la production de biomasse et le nombre de graines. Thèse de doctorat en Sciences agronomiques, ENSA de Montpellier, Montpellier, 103 pp.

Guilioni L., Lecoeur J. and Wéry J., 1998. Pois Protéagineux. Quel est l'impact des "coups de chaleurs"? *Persp. Agric.*, 236, 64–70.

Guilioni L., Wéry J. and Lecoeur J., 2003. High temperature and water deficit may reduce seed number in field pea purely by decreasing plant growth rate. *Functional Plant Biology*, 30, 1151–1164.

Guilioni L., Wéry J. and Tardieu F., 1997. Heat stress-induced abortion of buds and flowers in pea: is sensitivity linked to organ age or to relations between reproductive organs? *Ann. Bot.*, 80, 159–168.

Guinchard M.P., Robin C., Grieu P. and Guckert A., 1998. Cold acclimation in white clover subjected to chilling and frost: changes in water and carbohydrates status. *Eur. J. Agron.*, 6, 225–233.

Gusta L.V., Wilen R.W. and Fu P., 1996. Low temperature stress tolerance: the role of abscisic acid, sugars and heat-stable proteins. *Hortscience*, 31, 39–46.

Hamblin A.P. and Hamblin J., 1985. Root characteristics of some legume species and varieties on deep, free-draining entisols. *Aust. J. Agric Res.*, 36, 63–72.

Hamblin A.P. and Tennant D., 1987. Root length and water uptake in cereals and grain legumes : How well are they correlated ? *Aut. J. Agric. Res.*, 38: 513–527.

Hamon N., Bardner R. and Allen-Williams L., Lee J.B., 1987. Flight periodicity and infestation size of *Sitona lineatus*. *Ann. app. Biol.*, 111, 271–284.

Hardin S.C. and Sheehy J.E., 1980. Influence of shoot and root temperature on leaf growth, photosynthesis and nitrogen fixation of lucerne. *Ann. Bot.*, 45, 22–233.

Hardwick R.C., 1988. Les points critiques de la physiologie des protéagineux. *Persp. Agric.*, 121, 155–162.

Hare W.W. and Walker J.C., 1944. Ascochyta diseases of canning pea. *Wis. Agric. Exp. Stn. Res. Bull.*, 150, 1–31.

Harvey D., 1980. Seed production in leafless and conventional phenotypes of Pisum sativum L. in relation to water availability within a controled environment. *Ann. Bot.*, 45, 673–680.

Hauggaard-Nielsen H., Ambus P. and Jensen E.S. 2001. Interspecific competition, N use and interference with weeds in pea-barley intercropping. *Field Crops Res.*, 70, 101–109.

Hawkins R.C. and Cooper P.J.M., 1981. Growth, development and grain yield of maize. *Exp. Agr.*, 17, 203–207.

Healy W.E., Heins R.D. and Wilkins H.F., 1980. Influence of photoperiod and light quality on lateral branching and flowering of selected vegetatively-propagated plants. *J. Amer. Soc. Hort. Sci.*, 105 (6), 812–816.

Heath M.C. and Hebblethwaite P.D., 1985. Solar radiation interception by leafless, semi-leafless and leafed peas (*Pisum sativum*) under contrasting field conditions. *Ann. Appl. Biol.*, 107, 309–318.

Heath M.C. and Hebblethwaite P.D., 1987. Seasonal radiation interception, dry matter production and yield determination for a semi-leafless pea (*Pisum sativum*) breeding selection under contrasting field conditions. *Ann. Appl. Biol.*, 110, 413–420.

Heath M.C. and Wood R.K.S., 1971. Role of inhibitors of fungal growth in the limitation of leaf spots caused by Ascochyta pisi and Mycosphaerella pinodes. *Ann. Bot.*, 35, 475–491.

Hedley C.L., Lloyd J.R., Ambrose M.J. and Wang T.L., 1994. An analysis of seed development in *Pisum sativum* XVII. The effect of the rb locus alone and in combination with r on the growth and development of the seed. *Ann. Bot.*, 74, 365–371.

Herdina J.A. and Silsbury J.H., 1990. Growth, nitrogen accumulation and portioning and N₂ fixation in fababean (*Vicia faba*) and pea (*Pisum sativum* L.). *Field Crop Res.*, 24, 173–188.

Hirose T. and Werger M.J.A., 1987. Nitrogen use efficiency in instantaneous and daily photosynthesis of leaves in the canopy of a *Solidago altissima* stand. *Physiol. Plantarum*, 70, 215–222.

Hocking P.J. and Pate J.S., 1978. Accumulation and distribution of mineral element in annual lupin (*Lupinus albus* L. and *Lupinus angustifolius* L., *Aust J. Agric. Res.*, 29, 267–280.

Hole C.C. and Hardwick R.C., 1976. Development and control of the number of flowers per node in Pisum sativum L. Ann Bot., 40, 707–722.

Hole C.C. and Scott P.A., 1983. Effect of number and configuration of fruits, photon flux and age on the growth and dry matter distribution of fruits of *Pisum sativum* L. *Plant Cell Env.*, 6, 31–38.

Hurry V.M., Strand A., Tobiaeson M., Garderström P. and Öquist G., 1995. Cold hardening of spring and winter wheat and rape results in differential effects on growth, carbon metabolism, and carbohydrate content. *Plant Physiol.*, 109, 697–706.

Husain S.M. and Linck A.J., 1966. Relationship of apical dominance to the nutrient accumulation pattern in *Pisum sativum* var. alaska. *Physiol. Plant.*, 19, 992–1010.

Jensen E.S., 1986. The influence of rate and time of supply on nitrogen fixation and yield in pea (*Pisum sativum* L.). *Fertilizer Res.*, 10 , 193–202.

Jensen L.S., 1993. Effects of soil compaction on N mineralization and microbial C and N: field measurements and laboratory simulation. Nitrogen mineralization in agricultural soils, Haren (NL), ab-dlo.

Jeudy C., 2001. Effet de la date et de la densité de semis sur la mise en place de l'architecture du pois protéagineux de printemps (*Pisum sativum* L.). Perspectives pour la modélisation. Mémoire de fin d'études ENESAD, 47 p.

Jeuffroy M.H., 1987. Influence des hautes températures pendant la floraison sur l'élaboration du rendement du pois (*Pisum sativum* L.) cv. Solara. Diplôme d'Etude Approfondie, INA PG, Paris, 51 p.

Jeuffroy M.H., 1991. Pois. Les profils de graines: interprétation et modélisation. *Persp. Agric.*, 164, 62 to 72 p.

Jeuffroy M.H., 1991. Pois protéagineux et fortes températures. Perspectives Agricoles, 154, 87–93.

Jeuffroy M.H., 1991. Rôle de la vitesse de croissance, de la répartition des assimilats et de la nutrition azotée, dans l'élaboration du nombre de fraines du pois protéagineux de printemps (*Pisum sativum* L.). Thèse Université Paris XI, 208 p.

Jeuffroy M.H., 1994. Le nombre de graines par tige. In "Agrophysiologie du pois protéagineux; applications à la production agricole", INRA, ITCF, UNIP Eds, pp. 93–110.

Jeuffroy M.H. and Bouchard C., 1999. Intensity and duration of nitrogen deficiency on wheat grain number. *Crop Sci.*, 261, 1385–1393.

Jeuffroy M.H. and Chabanet C., 1994. A model to predict seed number per pod from early pod growth rate in pea (*Pisum sativum* L.). *J. Exp. Bot.*, 45, 709–715.

Jeuffroy M.H. and Devienne F., 1995. A simulation model for assimilate partitioning between pods in pea (Pisum sativum L.) during the period of seed set; validation in field conditions. *Field Crops Res.*, 41, 79–89.

Jeuffroy M.H. and Recous S., 1999. Azodyn : a simple model simulating the date of nirtogen deficiency for decision support in wheat fertilization. *Eur. J. Agron.*, 129–144.

Jeuffroy M.H., Duthion C., Meynard J.M. and Pigeaire A., 1990. Effect of a short period of high day temperatures during flowering on the seed number per pod of pea (*Pisum sativum* L.). *Agronomie*, 2, 139–145.

Jeuffroy M.H. and Ney B., 1997. Crop physiology and productivity. *Field Crops Res.*, 53, 3–16.

Jeuffroy M.H. and Sebillotte M., 1997. The end of flowering in pea: influence of plant nitrogen nutrition. *Eur. J. Agron.*, 6, 15–24.

Jeuffroy M.H. and Warembourg F.R., 1991. Carbon transfer and partitioning between vegetative and reproductive organs in Pisum sativum L. *Plant Physiol.*, 97, 440–448.

Jiang H.F. and Egli D.B., 1995. Soybean seed number and crop growth rate during flowering. *Agron. J.*, 87, 264–267.

Justes E., Jeuffroy M.H. and Mary B., 1997. Wheat, Barley and Durum Wheat; In: *Diagnosis of the Nitrogen Status in Crops*, Lemaire G. (Ed.), Springer, Berlin, 73–93.

Justes E., Mary B., Meynard J.M., Machet J.M. and Thelier-Huche L., 1994. Determination of a critical dilution curve for winter wheat crops. *Ann. Bot.*, 74, 397–407.

Knott C.M. and Belcher S.J., 1998. Optimum sowing dates and plant populations for winter peas (*Pisum sativum*). *J. Agric. Sci.*, 131, 449–454.

Kosslak R.M. and Bolhool B., 1984. Supression of nodule development of one side of split-root system of soybeans caused by prior inoculation of the other side. *Plant Physiol.*, 75, 125–130.

Kué J., 1995. Phytoalexin, stress metabolism and disease resistance in plants. *Ann. rev. Phytopathol.*, 33, 275–297.

Lambert R.G. and Linck A.J., 1958. Effects of high temperature on yield of peas. *Plant Physiol.*, 33, 347–350.

Landon F., Ferary S., Pierre D., Auger J., Biemont J.C., Levieux J. and Pouzat J., 1997. *Sitona lineatus* host-plant odors and their components: Effect on locomotor behavior and peripheral sensitivity variations. *J. Chem. Ecol.*, 23, 2161–2173.

Landon F., Levieux J., Huignard J., Rougon D. and Taupin P., 1995. Feeding activity of *sitona lineatus* L. on *Pisum sativum* L. during its imaginal life. *J. Appl. E ntomol.*, 119, 515–522.

Larmure A. and Munier-Jolain N.G., 2004. A crop model component simulating N partitioning during seed filling in pea. *Field Crops Res.*, 85, 135–148.

Lawrie A.C. and Wheeler C.T., 1974, The effect of flowering and fruit formation on the supply of photosynthetic assimilates to nodules of *Pisum sativum* L. in relation to the fixation of nitrogen. *New Phytol.*, 73, 1119–1127.

Lawrie A.C. and Wheeler C.T., 1974. The supply of photosynthetic assimilates to nodules of Pisum sativum L. in relation to the fixation of nitrogen. *New Phytol.*, 72, 1341–1348.

Layzell D.B., Pate J.S., Atkins C.A. and Canvin D.T., 1981. Partitioning of carbon and nitrogen and the nutrition of root and shoot apex in a nodulated legume. *Plant Physiol.*, 67, 30–36.

Le May C., 2002. Effet du couvert végétal du pois protéagineux sur le développement spatio-temporel de l'anthracnose à *Mycosphaerella pinodes*. Conséquences sur l'élaboration du rendement. Thèse de l'Ecoles Supérieure d'Agriculture de Rennes, 108 p.

Lecoeur J., 1994a. Réponses phénologiques et morphologiques du pois (Pisum sativum L.) à la contrainte hydrique. Conséquences sur le nombre de phytomères reproducteurs. Thèse de Doctorat Sciences Agronomiques, ENSAM, 99 p.

Lecoeur J., 1994b. Les indicateurs d'état hydrique du peuplement végétal: l'exemple du pois. Bulletin Semences FNAMS. 128, 11–14.

Lecoeur J. and Guilioni L., 1998. Rate of leaf production in response to soil water deficits in field pea. *Field Crops Res.*, 57, 319–328.

Lecoeur J. and Ney B., 2003. Change with time in potential radiation-use efficiency in field pea. *Eur. J. Agr.*, 19, 91–105.

Lecoeur J., Ney B. and Sinclair T.R. 2001. Which conceptual framework could be used to analyse the variability in yield of field pea? dans Proceedings of 3rd European Conferences on Grain Legumes. 8–12 July 2001. Cracow. Poland.

Lecoeur J. and Sinclair T.R., 1996. Field pea (*Pisum sativum* L.) transpiration and leaf growth in response to soil water deficits. *Crop Sci.*, 36, 331–335.

Lecoeur J. and Sinclair T.R., 2001a. Harvest index increase during seed growth of field pea. *Eur. J. Agron.*, 14, 173–180.

Lecoeur J. and Sinclair T.R., 2001b. Nitrogen accumulation, partitioning and nitrogen harvest index indrease during seed-fill of field pea. *Field Crops Res.*, 71, 87–99.

Lecoeur J. and Sinclair T.R., 2001c. Analysis of nitrogen partitioning in field pea plants resulting in linear increase in nitrogen harvest index. *Field Crops Res.*, 71, 151–158.

Lecoeur J., Wery J. and Sinclair T.R., 1996. Model of leaf area expansion on field pea (*Pisum sativum* L.) plants subjected to soil water deficits. *Agron. J.*, 88, 467–472.

Lecoeur J., Wery J., Turc O. and Tardieu F., 1995. Expansion of pea leaves subjected to short water deficit: Cell number and cell size are sensitive to stress at different periods of leaf development. *J. Exp. Bot.*, 46, 1093–1101.

Ledgard S.F., Morton R., Freney J.R., Bergersen F.J. and Simpson J.R., 1985. Assessment of the relative uptake of added and indigenous soil nitrogen by nodulated legumes and reference plants in the ^{15}N dilution measurement of N_2 fixation: derivation of method. *Soil Biol. Biochem.*, 17, 317–321.

Lejeune-Hénaut I., Bourion V., Eteve G., Cunot E., Delhaye K. and Desmyter C., 1999. Floral initiation in field-grown forage peas is delayed to a greater extent by short photoperiods, than in other types of European varieties. *Euphytica*, 109, 201–211.

Lejeune-Hénaut I., Hanocq E., Béthencourt L., Fontaine V., Delbreil B., Morin J., Petit A., Devaux R., Boilleau M., Stempniak J.J., Thomas M., Lainé A.L., Foucher F., Baranger A., Burstin J., Rameau C. and Giauffret C., 2008. The flowering locus *Hr* colocalizes with a major QTL affecting winter frost tolerance in *Pisum sativum* L. *Theor. Appl. Genet.*, 116: 1105–1116.

Lejeune-Hénaut I. and Wery J., 1994. Influence du froid sur la survie des plantes. In Agrophysiologie du pois protéagineux. Applications à la production agricole. INRA, ITCF and UNIP (Eds.), 280 p.

Lemaire G. and Allirand J.M., 1993. Relation entre croissance et qualité de luzerne : interaction génotype-mode d'exploitation. *Fourrages*, 134, 183–198.

Lemaire G., Cruz P., Gosse G. and Chartier M., 1985. Etude des relations entre la dynamique de prélèvement d'azote et la dynamique de croissance en matière sèche d'un peuplement de luzerne (*Medicago sativa* L.). *Agronomie* 5, 685–692.

Lemaire G. and Gastal F., 1997. N Uptake and Distribution in Plant Canopies. Diagnosis of the Nitrogen Status in Crops. G. Lemaire. Berlin, Springer, 3–43.

Lemaire G., Gastal F., Plenet D. and Le Bot. J., 1997. Le prélèvement de l'azote par les peuplements végétaux et la production des cultures. *In*: Maîtrise de l'azote dans les agrosystèmes. G. Lemaire, B. Nicolardot (Eds.) Série les colloques de l'INRA, INRA-Editions, Paris, pp. 121–139.

Lemaire G., Khaity M., Onillon B., Allirand J.M., Chartier M. and Gosse G., 1992. Dynamics of accumulation and partioning of N in leaves, stems and roots of lucerne (*Medicago sativa* L.) in a dense canopy. *Ann. Bot.*, 70, 429–435.

Le May C., Jumel S., Schoeny A. and Tivoli B., 2009b. Acochyta blight development on a new winter pea genotype highly reactive to photoperiod under field conditions. *Field Crop Research*, 111: 32–38.

Le May C., Ney B., Lemarchand E., Schoeny A. and Tivoli B., 2009b. Effect of pea plant architecture on the spatio-temporal epidemic development of ascochyta blight (*Mycosphaerella pinodes*) in the field. *Plant Pathology*, 58: 332–343.

Lemontey C., Mousset-Déclas C., Munier-Jolain N. and Boutin J.P., 2000. Endoreduplication level in pea seed is dependent on the maternal genotype and related to final seed size. *J. Exp. Bot.*, 51, 167–175.

Lemontey C., 1999. Influence du génotype maternel sur les divisions cellulaires dans l'embryon: conséquences pour le potentiel de croissance de la graine de pois. Thèse INA-PG. 97 p.

Lempereur B., 1996. Variabilité de caractères liés à l'adaptation climatique du pois protéagineux en semis d'automne. Rapport BTS Technologies végétales "Agronomie et systèmes de cultures". 35 pp.

Lepoivre P., 1982. Extraction d'ascochytine à partir des feuilles de pois infectées par Ascochyta pisi ou Mycosphaerella pinodes. *Parasitica*, 38 (2), 45–53.

Leterme P., 1985. Modélisation de la croissance et de la production des siliques chez le colza d'hiver (*Brassica napus* L.); application à l'interprétation de résultats de rendements. Thèse de Docteur-Ingénieur, INA-PG, Paris, 253 p.

Levitt J., 1980. Chilling, freezing and high temperature stress. *In*: Responses of plants to environmental stresses. Eds, Academic Press, p.

Lhomme J.P. and Guilioni L., 2004. A simple model for minimum crop temperature forecasting during nocturnal cooling. *Agric. Forest Meteorol.*, In Press, Corrected Proof.

Lhuillier-Soundele A., 1999. Analyse de l'élaboration de la teneur en azote des graines chez le pois (*Pisum sativum* L.). Modélisation de la répartition de l'azote vers les graines en remplissage. Thèse de doctorat de l'INA P-G. 116 p.

Lhuillier-Soundélé A., Munier-Jolain N.G. and Ney B., 1999a. Dependence of seed nitrogen concentration on plant nitrogen availability during the seed filling in pea. *Eur. J. Agron.*, 11, 157–166.

Lhuillier-Soundélé A., Munier-Jolain N.G. and Ney B., 1999b. Influence of nitrogen availability on seed nitrogen accumulation in pea. *Crop Sci.*, 39, 1741–1748.

Li P.H., 1994. Crop plant cold hardiness. In Physiology and determination of crop yield. American Society of Agronomy, Crop Science Society of America and Soil Science Society of America Eds, Madison, p.

Lineberger R.D. and Steponkus P.L., 1980. Cryoprotection by glucose, sucrose and raffinose to chloroplast thylakoids [Spinach leaves]. *Plant Physiol.*, 65, 298–304.

Livy Williams III, Schotzko D.J. and O'Keeffe L.E. 1995. Pea leaf weevil herbivory on pea seedlings: effects on growth response and yield. *Entomologia Experimentalis et Applicata*, 75, 255–269.

Lockhart J.A., 1965. An analysis of irreversible plant cell elongation. *J. Theor. Biol.*, 8, 264–275.

Lovell L.P.H. and Lovell P.J., 1970. Fixation of CO_2 and export of photosynthate by the carpel in Pisum sativum. *Physiol. Plant.*, 23, 316–322.

Madeira A.C. and Clark J.A., 1995. The principles of ressource captures in relation to necrotrophic infection. *Asp. App. Biol.*, 42, 19–31.

Maltas A., 2002. Mise au point et évaluation d'un modèle pour déterminer a priori les conséquences agronomiques et environnementales de l'insertion de nouvelles variétés de pois d'hiver. Mémoire de DEA, Grignon, UMR d'Agronomie: 22 + annexes.

Mangin M., 1998. Pois protéagineux quels sont les risques liés au climat ? *Persp. Agric.*, 239, 78–82.

Manichon H. and Gautronneau Y., 1987. Guide méthodologique du profil cultural.

Martin I., Tenorio J.L. and Ayerbe L. 1994. Yield, growth, and water-use of conventional and semileafless peas in semiarid environnements. *Crop Sci.*, 34(6), 1576–1583.

Matile P., 1992. Chloroplast senescence. In Baker N.R., Thoms H. (ed). Crop photosynthesis: spatial and temporal determinant. Amsterdam Elsevier, 413–440.

Maurer A.R., Jaffray D.E. and Fletcher H.F., 1966. Response of peas to environment. III. Assessment of the morphological development of peas. *Can. J. Plant Sci.*, 46, 285–290.

Mazliak P., 1992. Les effets du froid sur les biomembranes. *In*: Les végétaux et le froid. D. Côme Eds, Hermann, Paris, 600 p.

Mc Neil D.L., Atkins C.A. and Pate J.S., 1979. Uptake and utilization of xylem-borne amino compounds by shoot organs of a legume. *Plant Physiol.*, 63, 1076–1081.

Meckel L., Egli D.B., Phillips R.E., Radcliffe D. and Leggett J.E., 1984. Effect of moisture stress on seed growth in soybeans. *Agron. J.*, 76, 647–650.

Mengel K., 1994. Symbiotic dinitrogen fixation—Its dependance on plant nutrition and its ecophysiological impact. *Z. Pflanzenernähn*, 157, 233–241.

Mériaux B., 2002. Semences de pois : des « trucs » pour récolter à la bonne teneur en eau. *Bull. Semences*, 166, 21–22.

Meynard J.M., 1985. Construction d'itinéraires techniques pour la conduite du blé d'hiver. Paris, INA PG, 297.

Meynard J. M. and Doré T., 1992. Agronomic diagnosis and cropping systems improvements. 2nd congress of the European Society of Agronomy, Warwick.

Miller D.G., Manning C.E. and Teare I.D., 1977. Effects of soil water levels on components of growth and yield in peas. *J. Am. Soc. Hort. Sci.*, 102, 349–351.

Mitchell R.L. and Russel W.J., 1971. Root developpement and rooting patterns of soybean (*Glycine max* L. Merril) evaluated under field conditions. *Agron. J.*, 63, 313–316.

Monteith J.L., 1972. Solar radiation and productivity in tropical ecosystems. *J. Appl. Ecol.*, 9, 747–766.

Monteith J., 1977. Climate and the efficiency of crop production in Britain. *Philos. T. Roy. Soc. B*, 281, 277–294.

Muchow R.C. and Carberry P.S. 1990. Phenology and leaf area development in a tropical grain sorghum. *Field Crops Res.*, 23, 221–237.

Muller B. and Touraine B., 1992. Inhibition of NO_3^- uptake by various phloem translocated amino acids in soybean seedlings. *J. Exp. Bot.*, 43, 617–623.

Munier-Jolain N.M., 1996. Analyse et modélisation de la structure du peuplement et de la production de graines chez le lupin blanc (*Lupinus albus* L.) de printemps. Effets de la date et de la densité de semis. *Thèse INAPG*, 156 p.

Munier-Jolain N.G. and Carrouée B., 2003. Quelle place pour le pois dans une agriculture respectueuse de l'environnement: argumentaire agri-environnemental. *Cahiers Agriculture*, 12, 111–120.

Munier-Jolain N.G., Munier-Jolain N.M., Roche R., Ney B. and Duthion C., 1998. Seed growth rate in grain legumes. I: seed growth is not affected by assimilates availability on seed growth rate. *J. Exp. Bot*, 49, 1963–1969.

Munier-Jolain N.G. and Ney B., 1998. Seed growth rate in grain legumes. II: Seed growth rate depends on cotyledon cell number. *J. Exp. Bot*, 49, 1971–1976.

Munier-Jolain N.G., Ney B. and Duthion C., 1993. Sequential development of flowers and seeds on the mainstem of an indeterminate soybean. *Crop Sci.*, 33, 768–771.

Munier-Jolain N.G., Ney B. and Duthion C., 1996. Termination of seed growth in relation to nitrogen content of vegetative parts in soybean plants. *Eur. J. Agr.*, 5, 219–225.

Munier-Jolain N.M., Ney B. and Duthion C., 1997. Analysis of sequential reproductive development in white lupin cv. Lublanc. *Aust. J. Agric. Res.*, 48, 913–922.

Munier-Jolain N.G. and Salon C., 2003. Can sucrose content in phloem sap reaching field pea seeds (*Pisum sativum* L.) be a accurate indicator of seed growth potential? *J. Exp. Bot.*, 54, 392, 2457–2465.

Murfet I.C., 1971. Flowering in *Pisum*. A three-gene system. *Heredity*, 27, 93–110.

Murfet I.C., 1971. Flowering in *Pisum*. Reciprocal grafts between known genotypes. *Aust. J. Biol. Sci.*, 24, 1089–1101.

Murfet I.C., 1973. Flowering in *Pisum*. *Hr*, a gene for high response to photoperiod. *Heredity*, 26, 243–257.

Murfet I.C., 1975. Flowering in *Pisum*: multiple alleles at the *lf* locus. *Heredity*, 35, 85–98.

Murfet I.C., 1982. Flowering in garden pea: expression of gene SN in field and use of mulyiple characters to detect segregation. *Crop Sci.*, 22, 923–926.

Murfet I.C., 1985. Flowering in *Pisum*: a sixth locus, *Dne*. *Ann. Bot.*, 56, 835–846.

Murfet I.C. and Reid J.B., 1993. Developmental mutants. *In* Peas: Genetics, Molecular Biology and Biotechnology. R. Casey and D.R. Davies eds, CAB International, Wallingford, UK, 323 pp. 165-216.

Nagao M.A. and Rubinstein B., 1976. Early events associated with lateral bud growth of *Pisum sativum* L. *Bot. Gaz.*, 137, 39–44.

Neo H.H. and Layzell D.B., 1997. Phloem glutamine and the regulation of O_2 diffusion in legume nodules. *Plant Physiol.*, 113, 259–267.

Ney B., 1994. Seed number per stem of pea is linearly related to its growth rate between flowering and beginning of seed filling. C.R. activité du programme Agro-Industry-Research, pp. 1–4.

Ney B., 1994. Modélisation de la croissance aérienne d'un peuplement de pois. Agrophysiologie du pois protéagineux, UNIP-ITCF-INRA, mai 1994, 39–48.

Ney B. and Duc G., 1996. Potential and problems with winter sowing of food legumes in northern Europe. AEP Workshop. Dijon 3–4 décembre 1996. 35–42.

Ney B. and Turc O. 1993. A heat unit description of the reproductive development of pea (*Pisum sativum* L.). *Crop Sci.*, 33, 267–70.

Ney B. and Turc O., 1993. Heat-unit based description of the reproductive development of pea. *Crop Sci.*, 33, 510–514.

Ney B. and Turc O., 1994. Modélisation du développement reproducteur du pois en conditions non limitantes. In Ney B., Duchêne E., Carrouée B., Angevin F. (eds). Agrophysiologie du pois protéagineux : applications à la production agricole. INRA- ITCF- UNIP, Paris, 280, 27–34 p.

Ney B., Doré T. and Sagan M., 1997. The nitrogen requirement of major agricultural crops: Grain Legumes. *In*: Diagnosis of the nitrogen status in crops. G. Lemaire (Ed.) Springer-Verlag, Heigelberg, pp. 107–118.

Ney B., Duthion C. and Fontaine E., 1993. Timing of reproductive abortions in relation to cell division, water content and growth of the seeds of pea seeds. *Crop Sci.*, 33, 267–270.

Ney B., Duthion C. and Turc O., 1994. Phenological response of pea to water stress during reproductive development. *Crop Sci.*, 34, 141–146.

Nougarède A. and Rembur J., 1985. Le point végétatif en tant que modèle pour l'étude du cycle cellulaire et de ses points de contrôle. Bulletin de la Société Botanique Française, Actualité Botanique, 132, 9–34.

Nougarède A. and Rondet P., 1973a. Les stipules du Pisum sativum L. var. nain hatif d'Annonay et leurs relations avec la feuille à l'état jeune. Compte Rendu de l'Académie des Sciences de Paris, série D., 277, 393–396.

Nougarède A. and Rondet P., 1973b. Un modèle original d'organisation de la tige: étude du fonctionnement plastochronique chez le Pisum sativum L. var. nain hâtif d'Annonay. Comptes Rendus de l'Académie des Sciences de Paris, série D., 277, 997–1000.

Nougarède A. and Rondet P., 1978. Evènements structuraux et métaboliques dans les entre-nœuds des bourgeons axillaires du pois, en réponse à la levée de dominance. *Can. J. Bot.*, 56, 1213–1228.

Passioura J.B., 1996. Simulation models: science, snake oil, education, or engineering? *Agron. J.*, 88, 690–694.

Pate J.S., 1958. Nodulation studies in legumes. I. The synchronisation of host and symbiotic development in the field pea, *Pisum Arvense* L. *Aust. J. Biol. Sci.*, 11, 361–381.

Pate J.S., 1975. Pea. *In*: Crop physiology: some cases histories, Evans L.T. (Ed.), Cambridge university Press, 191–224.

Pate J.S., 1980. Transport and partitioning of nitrogenous solutes. *Ann. Rev. Plant Phys.*, 31, 313–340.

Pate J.S. and Atkins C.A., 1983. Xylem transport and the functional economy of carbon and nitrogen of a legume leaf. *Plant Physiol.* 71, 835–840.

Pate J.S., Atkins C.A., Hamel K., McNeil D.L. and Layzell D.B., 1979a. Transport of organic solutes in phloem and xylem of a nodulated legume. *Plant Physiol.*, 63, 1082–1088.

Pate, J.S., Atkins C.A., Herridge D.F. and Layzell D.B., 1981. Synthesis, storage and utilization of amino compounds in white lupin (*Lupinus albus* L.). *Plant Physiol.*, 67, 37–42.

Pate J.S. and Flinn A.M., 1973. Carbon and nitrogen transfer from vegetative organs to ripening seeds of field Pea (*Pisum arvense* L.). *J. Exp. Bot.*, 24, 1090–1099.

Pate J.S. and Herridge D.F., 1978. Partitioning and utilization of net photosynthate in a nodulated legume. *J. Exp. Bot.*, 29 (109), 401–412.

Pate J.S. and Layzell D.B., 1981. Carbon and nitrogen partitioning in the whole plant. A thesis based on empirical modeling. In: Bedley JD, ed. *Nitrogen and carbon metabolism*, Martinus Nijhoff / Dr W Junk Publishers, The Hague—Boston—London, pp. 95–134.

Pate J.S., Layzell D.B. and McNeil D.L., 1979b. Modelling the transport and utilization of carbon and nitrogen in a nodulated legume. *Plant Physiol.*, 63, 730–737.

Pate J.S., Sharkey P.J. and Atkins C.A., 1977. Nutrition of a developing legume fruit. Functional economy in terms of carbon and nitrogen in a nodulated legume. *Plant Physiol.*, 59, 506–510.

Pate J.S., Sharkey P.J. and Lewis O.A.M., 1975. Xylem to phloem transfer of solutes in fruiting shoots of legumes, studied by a phloem bleeding technique. *Planta*, 122, 11–26.

Paton D.M., 1968. Photoperiodic and temperature control of flower initiation in the late pea cultivar greenfeast. *Aust. J. biol. Sci.*, 21, 609–617.

Pearce R.S., 1999. Molecular analysis of acclimation to cold. *Plant Growth Regul.*, 29, 47–76.

Peoples M.B. and Dalling M.J., 1988. The interplay between proteolysis and amino acids metabolism during senescence and nitrogen reallocation. In: Senescence and aging in plants, Nooden L.D., Leopold A.C. (ed). Academic Press, NY, 182–217.

Peoples M.B., Pate J.S. and Atkins C.A., 1983. Mobilization of nitrogen in fruiting plants of a cultivr of cowpea. *J. Exp. Bot.*, 34, 142, 563–578.

Peoples M.B., Pate J.S., Atkins C.A. and Murray D.R., 1985. Economy of water, carbon and nitrogen in the developing cowpea fruit. *Plant Physiol.*, 77, 142–147.

Pic E., 1998. Analyse de la senescence foliaire chez le pois (Pisum sativum L., cv Messire) : utilisation conjointe d'un modèle de développement et de marqueurs moléculaires. Thèse ENSAM. 159 p.

Pic E., Teyssendier de la Serve B., Tardieu F. and Turc O. 2002. Leaf senescence induced by mild water deficit follows the same sequence of macroscopic, biochemical, and molecular events as monocarpic senescence in pea. *Plant Physiol.*, 128, 236–246.

Pierce M. and Bauer W.D., 1983. A rapid regulatory response governing nodulation in soybean. *Plant Physiol.*, 73, 286–290.

Pigeaire A., 1986. Relation entre le nombre de bourgeons végétatifs et le nombre de ramifications par plante chez le soja de type indéterminé. *Informations Techniques CETIOM*, 95, 3–7.

Pigeaire A., Duthion C. and Turc O., 1986. Characterisation of final stage of seed abortion in indetreminate soybean, white lupin and pea. *Agronomie*, 6, 371–378.

Plénet D. and Cruz P., 1997. Maize and Sorghum. Diagnosis of the Nitrogen Status in Crops. G. Lemaire. Berlin, Springer, 93–106.

Powell A., 1985. Impaired membrane integrity—A fundamental cause of seed-quality differences in peas ; In *The pea crop: a basis for improvement*, Butterworths ed. London. 383–394 p.

Pueppke S.G., 1983. Rhizobium infection thread in root hairs of *Glycine max* (L.) Merr., *Glycine Soja* Sieb and Zucc., and *Vigna unguiculata* (L.) Walp. *Can. J. Microbiol.*, 29, 69–76.

Pumphrey F.V., Ramig R.E. and Allmaras R.R., 1979. Field response of peas (*Pisum sativum* L.) to precipitation and excess heat. Journal of American Society of Horticultural Science, 104 (4), 548–550.

Pyke K.A. and Hedley C.L., 1985. Growth and photosynthesis of different pea phenotypes. *In*: P.D. Hebblethwaite, M.C. Heath and T.C.K. Dawkins (Editors), The pea crop. Butterworths, London, 297–305 p.

Ragonnaud G., 2001. Stabilité du rendement et de la teneur en protéines du pois protéagineux. Etude des interactions génotype—environnement. Mémoire de fin d'études INA-PG, 38 p.+annexes.

Rameau C., Dénoue D., Fraval F., Haurogné K., Josserand J., Laucou V., Batge S. and Murfet I.C., 1998. Genetic mapping in pea. 2. Identification of RAPD and SCAR markers linked to genes affecting plant architecture. *Theor. Appl. Genet.*, 97: 916–928.

Ratliff L.F., Ritchie J.T. and Cassel D.K., 1983. Field-Measured Limits of Soil Water Availability as related to laboratory-measured properties. *Soil Sci. Soc. Am. J.*, 47, 770–775.

Ravn H.P. and Jensen E.S., 1992. Effect of pea and bean weevil (*Sitona lineatus* L.) on N_2-fixation and yield in peas. 1st European Conference on Grain Legumes, Angers, France, 347–348.

Rennie R.J. and Rennie D.A., 1983. Techniques for quantifying N_2 fixation in association with nonlegumes under field and greenhouse conditions. *Can. J. Microbiol.*, 29 (8), 1022–1035.

Rey H., Guilioni L. and Lecoeur J., 2001. 3D numerical pea (Pisum sativum L.) plants responding to temperature. dans Actes de 3rd European Conferences on Grain Legumes. 8–12 July 2001. Cracovie, Pologne.

Ridao E., Oliveira C.F., Conde J.R. and Minguez M.I., 1996. Radiation interception and use, and spectral reflectance of contrasting canopies of autumn sown faba beans and semi-leafless peas. *Agric. Forest Meteorol.*, 79 (3), 183–203.

Roche R., 1998. Modélisation de la variabilité des profils de graines associée à la séquentialité du développement reproducteur chez différents génotypes de pois protéagineux. Thèse de doctorat INA-PG. 101 p.

Roche R. and Jeuffroy M.H., 2000. A model to calculate the vertical distribution of grain number in pea. *Agron. J.*, 92, 663–671.

Roche R., Jeuffroy M.H. and Ney B., 1998. A model to simulate the final number of reproductive nodes in pea (*Pisum sativum* L.). *Ann. Bot.*, 81, 545–555.

Roche R., Jeuffroy M.H. and Ney B., 1999. Comparison of different models predicting the date of beginning of flowering in pea (*Pisum sativum* L.). *Ecol. Modelling*, 118, 213–226.

Roger C. and Tivoli B., 1996. Spatio-temporal development of pycnidia and perithecia and dissemination of spores of *Mycosphaerella pinodes* on pea (*Pisum sativum*). *Plant Pathol*, 45, 518–528.

Roger C., Tivoli B. and Huber L., 1999. Effects of temperature and moisture on disease and fruiting body development of *Mycosphaerella pinodes* (BerK; and Blox.) on pea. *Plant Pathol.*, 48, 1–9.

Roger C., Tivoli B. and Lemarchand E., 1998. Epidémiologie de l'anthracnose. *Phytoma*, 509, 32–36.

Rubinstein B. and Nagao M.A., 1976. Lateral bud outgrowth and its control by the apex. *Bot. rev.*, 42,83–113.

Ruelland E., Vaultier M.-N., Zachowski A. and Hurry V., 2009. Cold Signalling and Cold Acclimation in Plants. Advances in Botanical Research, 49, Book Series: Advances in Botanical Research incorporating Advances in Plant Pathology, 49, 35–150.

Sagan M., 1993. La symbiose Rhizobium-légumineuses: analyse de mutants symbiotiques de pois *Pisum sativum* L. Thèse de doctorat de l'Université Paris XI Orsay, 125 p.

Sagan M. and Gresshoff P.M., 1996. Developmental mapping of nodulation events in pea (Pisum sativum L.) using supernodulating plant genotypes and bacterial variability reveals both plant and Rhyzobium control of nodulation regulation. *Plant Sci.*, 117, 167–179.

Sagan M., Ney B. and Duc G., 1993. Plant symbiotic mutants as a tool to analyse nitrogen and yield relationship in field-grown peas (*Pisum sativum L.*). *Plant Soil*, 153, 33–45.

Salette J. and Lemaire G., 1981. Sur la variation de la teneur en azote des graminées fourragères pendant leur croissance: formulation d'une loi de dilution. *CR Ac. Sc. Paris*, 292, 875–878.

Salon S., Munier-Jolain N.G., Duc G., Voisin A.S., Grandgirard D., Larmure A., Emery R.J.N. and Ney B., 2001. Grain legume seed filling in relation to nitrogen acquisition: a review and prospects with particular reference to pea. *Agronomie*, 21, 539–552.

Salter P.J., 1962. Some responses of peas to irrigation at different growth stages. *J. Hort. Sci.*, 37, 141–149.

Salter P.J. and Drew D.H., 1965. Root growth as a factor in the response *of Pisum sativum* L. to irrigation. *Nature*, 4988, 1063–1064.

Sarr B., Lecoeur J. and Clouvel P., 2004. Scheduling irrigation using water balance model: application to irrigation methods researches for confectionery groundnut (*Arachis hypogeaea* L.) quality in Senegal. *Agr. Water Manage.*, sous presse.

Schjoerring J.K. and Mattsson M., 2001. Quantification of ammonia exchange between agricultural cropland and the atmosphere: measurements over two complete growth cycles of oilseed rape, wheat, barley and pea. *Plant Soil*, 228, 105–115.

Schoeny A., Menat J., Darsonval A., Rouault F., Jumel S. and Tivoli B., 2008. Effect of pea canopy architecture on splash dispersal of *Mycosphaerella pinodes* conidia. Plant Pathology 57: 1073–1085.

Scholes J.D., 1992. Photosynthesis: cellular and tissue aspects in disease leaves. p 85–106. In: Pest and Pathogens: Plant responses to foliar attack. PC Ayres, ed Bios Scientific Publishers, Oxford.

Schotzko D.J. and Quisenberry S.S., 1999. Pea leaf weevil (Coleoptera: Curculionidae) spatial distribution in peas. *Environ. Entomol.*, 28, 477–484.

Serraj R., Sinclair T.R. and Purcell L.C., 1999. Symbiotic N_2 fixation response to drought. *J. Exp. Bot.*, 50, 143–155.

Silsbury H.J.H., 1990. Growth, nitrogen accumulation and portioning, and N2 fixation in faba bean (*Vicia faba* cv. Fiord) and pea (*Pisum sativum* L cv Early Dun). *Field Crops Res.*, 24, 173–188.

Sinclair T.R., 1994. Limits to crop yield. In 'Physiology and determination of crop yield'. Eds B.K.J., J.M. Bennett, T.R. Sinclair and Paulsen G.M. ASA, CSSA, SSSA, Madison, WI, USA.

Sinclair T.R. and de Witt C.T., 1976. Analysis of the carbon and nitrogen limitations to soybean yield. *Agron. J.*, 68, 319–324.

Sinclair T.R. and Horie, 1989. Leaf nitrogen, photosynthesis, and crop radiation use efficiency: a review. *Crop Sci.*, 29, 90–98.

Sinclair T.R. and Muchow C., 1999. Radiation use efficiency. *Adv. Agron.*, 65, 215–263.

Sinclair T.R., Salado-Navarro L., Morandi E.N., Bodrero M.L. and Martignone R.A., 1992. Soybean yield in Argentina in response to weather variation among cropping seasons. *Field Crops Res.*, 30, 1–11.

Soussana J.F. and Hartwig U.A., 1996. The effect of elevated CO_2 on symbiotic N_2 fixation: a link between the carbon and nitrogen cycles in grass land ecosystems. *Plant Soil*, 187 (2), 321–332.

Sparrow S.D. and Cochran V.L., Sparrow E.B., 1995. Dinitrogen fixation by seven legume crops in Alaska. *Agron. J.*, 87, 34–41.

Spiertz J.H.J., 1977. The influence of temperature and light intensity on grain growth in relation to the carbohydrate and nitrogen economy of the wheat plant. *Neth. J. Agric. Sci.*, 25, 182–197.

Sprent J.I., Stephens J.H. and Rupela O.P., 1988. Environmental effects on nitrogen fixation. *In:* World crops : cool season food legumes. R.J. Sumerfield (Ed.) ISBN 90-247-3641-2.

Sprent J.I. and Thomas R.J., 1984. Nitrogen nutrition of seedling grain legumes : some taxonomis, morphological and physiological constraints. *Plant Cell Env.*, 7, 637–645.

Stanfield B., Ormrod D.P. and Fletcher H F., 1966. Response of peas to environment. II. Effects of temperature in controlled-environment cabinets. *Can. J. Plant Sci.*, 46, 195–203.

Starling M.E., Wood C.W., Weaver D.B., 1998. Starter nitrogen and growth habit effects on late-planted soybean. *Agron. J.*, 90, 658–662.

Steene F., Vulsteke, G., 1999. Influence of the sowing date and the cultivar on the appearance and density of the pea weevil, *Sitona lineatus* (L.) in pea, *Pisum sativum* (L.) crops. *Parasitica*, 55, 195–202.

Stocker R., 1973. Response of viner peas to water during different phases of growth. *N. Z. J. Exp. Agri.*, 1, 73–76.

Streeter J.G., 1988. Inhibition of legume nodule formation and N_2 fixation by nitrate. *CRC Crit. Rev. Plant Sci.*, 7, 1–23.

Summerfield R.J. and Roberts E.H., 1988. Photothermal regulation of flowering in pea, lentil, faba bean and chickpea. In World crops: cool season food legumes. Summerfield R.J., Kluwer Academic Publishers, Amsterdam.

Taji T., Ohsumi C., Iuchi S., Seki M., Kasuga M., Kobayashi M., Yamaguchi-Shinozaki K. and Shinozaki K., 2002. Important roles of drought- and cold-inducible genes for galactinol synthase in stress tolerance in *Arabidopsis thaliana*. *Plant J.*, 29, 417–426.

Tardieu F. and Manichon H., 1987. Etat structural, enracinement et alimentation hydrique du maïs. II.-Croissance et disposition spatiale du système racinaire. *Agronomie*, 7 (3), 201–211.

Taupin P., Bournoville R. and Thieuleux J., 1994. Stress biotiques dus aux ravageurs. In Ney B., Duchêne E., Carrouée B., Angevin F. (Eds.). Agrophysiologie du pois protéagineux : applications à la production agricole. INRA- ITCF-UNIP, Paris, 280 p., 221–227.

Thibaud N., 2004. Récolte des semences de pois et féverole : approche comparative des équipements et réglages. *Bull. Semences*, 178. Sous presse.

Thorup-Kristensen K., 1998, Root growth of green pea (Pisum sativum L.) genotypes. *Crop Sci.*, 38, 1445–1451.

Tivoli B., 1994. Conséquences des attaques parasitaires foliaires sur l'élaboration du rendement des plantes à croissance indéterminée. In Agrophysiologie du pois protéagineux. Applications à la production agricole. Paris, mai 1994. Ed.INRA/ITCF/UNIP, Paris, 199–219.

Tivoli B., Béasse C., Lemarchand E. and Masson E., 1996. Effect of Ascochyta blight (*Mycosphaerella pinodes*) on yield components of single pea (*Pisum sativum*) plants under field conditions. *Ann. Appl. Biol.*, 129, 207–216.

Tivoli B. and Samson R., 1996. Biotic constraints on yield and produce quality. In Problems and prospects for winter sowing of grain legumes in Europe. AEP Eds, Paris, 119 p.

Townley-Smith L. and Wright A. T., 1994. Field pea cultivar and weed response to crop seed rate in western Canada. *Can. J. Plant Sci.*, 76, 900–906.

Tricot F., 1993. Mise en place des nodosités du pois protéagineux de printemps (Pisum Sativum L.) Influence de la nutrition carbonée. Thèse de Doctorat, Université Paris-Sud Orsay, France, 78 p.

Tricot F., Crozat Y. and Pellerin S., 1997. Root growth and nodule establishment on pea (*Pisuma sativum* L.). *J. Exp. Bot.*, 48, 316, 1935–1941.

Tricot-Pellerin F., Angevin F. and Crozat Y., 1994. Elaboration de la biomasse des nodosités: influence de la nutrition carbonée. In Ney B., Duchêne E., Carrouée B., Angevin F. (Eds.). Agrophysiologie du pois protéagineux: applications à la production agricole. INRA-ITCF-UNIP, Paris, 280, 75–91 p.

Truong H.H. and Duthion C., 1993. Time of flowering of pea (*Pisum sativum* L.) as a function of leaf appearance rate and node of first flower. *Ann. Bot.*, 72, 133–142.

Turc O., 1988. Elaboration du nombre de graines chez le pois protéagineux (*Pisum sativum* L. cv. Frisson, Finale et leurs homolgues afila): influence du rayonnement intercepté et application au diagnostic cultural. Thèse Université des Sciences et Techniques du Languedoc, Montpellier.

Turc O., Farinha N.C., Sousa C.J.F. and Ney B., 1994. A characterization of seed growth and water content to describe the development of chickpeas seeds. In M. Borin et M. Sattin (Eds.). Proc. 3nd ESA Congress. 18–22 September 1994. Abano-Padova. Italie.

Turc O. and Lecoeur J., 1997. Leaf primordium initiation and expanded leaf production are co ordinated through similar response to air temperature in pea (*Pisum sativum* L.). *Ann. Bot.*, 80, 265–273.

Turc O., Wery J. and Sao Chan Cheong G. 1990. A limited drought stress as a tool for the management of the balance between vegetative and reproductive component of peas (Pisum sativum L.). dans Proceeding of the first congress of the European Society of Agronomy, 5–7 décembre, Paris.

Van den Heuvel J. and Grootvel D, 1980. Formation of phytoalexins within and outside lesions of Botrytis cinerea in french bean leaves. *Neith. J. Plant Pathol.*, 86, 27–35.

Varlet-Grancher C., Gosse G., Chartier M., Sinoquet H., Bonhomme R. and Allirand J.M. 1989. Mise au point: rayonnement solaire absorbé ou intercepté par un couvert végétal. *Agronomie*, 9, 419–439.

Vessey J.K., 1992. Cultivar differences in assimilate partitioning and capacity to maintain N2 fixation rate in pea during pod-filling. *Plant Soil*, 139, 185–194.

Voisin A.S., 2002. Etude du fonctionnement des racines nodulées du pois (Pisum sativum L.) en relation avec la disponibilité en nitrates du sol, les flux de carbone au sein de la plante rt la phénologie. Dijon, Bourgogne, 129 p.

Voisin A.S., Salon C., Jeudy C. and Warembourg F.R., 2003a. Seasonal patterns of ^{13}C partitionning between shoot and nodulated roots of N_2^- or nitrate fed *Pisum sativum* L. *Ann. Bot.*, 91, 539–546.

Voisin A.S., Salon C., Jeudy C. and Warembourg F.R., 2003b. Root and nodule growth in *Pisum sativum* L. in relation to photosynthesis. Analysis using ^{13}C labelling, *Ann. Bot.*, 92, 1–7.

Voisin A.S., Salon C., Jeudy C. and Warembourg F.R., 2003c. Symbiotic N_2 fixation in relation to C economy of *Pisum sativum* L. as a funcion of plant phenology, *J. Exp. Bot.*, 54 (393), 2733–2744.

Voisin A.S., Salon C., Munier-Jolain N.G. and Ney B., 2002a. Effect of mineral nitrogen on nitrogen nutrition and biomass partitioning between the shoot and roots of pea (*Pisum sativum* L.). *Plant Soil*, 242, 251–262.

Voisin A.S., Salon C., Munier-Jolain N.G. and Ney B., 2002b. Quantitative effect of soil nitrate, growth potential and phenology on symbiotic nitrogen fixation of pea (*Pisum sativum* L.). *Plant Soil*, 243, 31–42.

Wall D.A., Friesen G.H. and Bhati T.K., 1991. Wild mustard interference in traditional and semi-leafless peas. *Can. J. Plant Sci.*, 71, 473–480.

Wall D.A. and Townley-Smith L. 1996. Wild mustard (*Sinapis arvensis*) response to field pea (*Pisum sativum*) cultivar and seeding rate. *Can. J. Plant Sci.*, 76, 907–914.

Wang T.L. and Hedley C.L., 1993. Genetic and developmental analysis of the seed. In Peas: genetics, molecular biology andbiotechnology. Casey R., Davies D.R. (Eds.). CAB international, UK. 83–120.

Wanner L.A. and Juntilla O., 1999. Cold-induced freezing tolerance in *Arabidopsis. Plant. Physiol.*, 120: 391–399.

Warembourg F.R., Haegel B., Fernandez M. and Montange D., 1984. Distribution et utilisation des assimilates carbonés en relation avec la fixation symbiotique d'azote chez le soja (*Glycine max* L. Merrill). *Plant Soil*, 82, 163–178.

Weeden N.F., Ellis T.H.N., Timmerman-Vaughan G.M., Swiecicki W.K., Rozov S.M. and Berdnikov V.A., 1998. A consensus linkage map for Pisum sativum. *Pisum Genet.*, 30: 1–4.

Weller J.L., Reid J.B., Taylor S.A. and Murfet I.C., 1997. The genetic control of flowering in pea. *Trends Plant Sci.*, 2, 412–418.

Wéry J., 1987. Relations entre la nutrition azotée et la production chez les légumineuses. Nutrition azotée des légumineuses. P. Guy, Les Colloques de l'INRA, 37, 199–223.

Wright D.P., Baldwin B.C., Shephard M.C. and Scholes J.D., 1995. Source sink relationships in wheat leaves infected with powdery mildew. II. Changes in the regulation of the Calvin cycle. *Physiol. Mol. Plant Pathol.*, 47, 255–267.

Yordanov I., Georgieva K., Tsonev T. and Velikova V., 1996. Effect of cold hardening on some photosynthetic characteristics of pea (Pisum sativum L. cn Ran1) plants. *Bulgarian J. Plant Physiol.*, 22, 13–21.

Zuther E., Büchel K., Hundertmark M., Stitt M., Hincha D.K. and Heyer A.G., 2004. The role of raffinose in the cold acclimation response of *Arabidopsis thaliana*. FEBS Lett., 576: 169–173.

Index